Memoirs
of the
American Mathematical Society

Number 979

Erdős Space and Homeomorphism Groups of Manifolds

Jan J. Dijkstra
Jan van Mill

November 2010 • Volume 208 • Number 979 (fourth of 6 numbers) • ISSN 0065-9266

American Mathematical Society
Providence, Rhode Island

Library of Congress Cataloging-in-Publication Data

Dijkstra, Jan J. (Jan Jakobus), 1953-
  Erdős space and homeomorphism groups of manifolds / Jan J. Dijkstra, Jan van Mill.
    p. cm. — (Memoirs of the American Mathematical Society, ISSN 0065-9266 ; no. 979)
  "November 2010, Volume 208, number 979 (fourth of 6 numbers)."
  Includes bibliographical references and index.
  ISBN 978-0-8218-4635-3 (alk. paper)
  1. Homeomorphisms. 2. H-spaces. 3. Topological groups. I. Mill, Jan van, 1951- II. Title.
QA614.D55   2010
514—dc22   2010030133

Memoirs of the American Mathematical Society

This journal is devoted entirely to research in pure and applied mathematics.

Publisher Item Identifier. The Publisher Item Identifier (PII) appears as a footnote on the Abstract page of each article. This alphanumeric string of characters uniquely identifies each article and can be used for future cataloguing, searching, and electronic retrieval.

Subscription information. Beginning with the January 2010 issue, *Memoirs* is accessible from www.ams.org/journals. The 2010 subscription begins with volume 203 and consists of six mailings, each containing one or more numbers. Subscription prices are as follows: for paper delivery, US$709 list, US$567 institutional member; for electronic delivery, US$638 list, US$510 institutional member. Upon request, subscribers to paper delivery of this journal are also entitled to receive electronic delivery. If ordering the paper version, subscribers outside the United States and India must pay a postage surcharge of US$65; subscribers in India must pay a postage surcharge of US$95. Expedited delivery to destinations in North America US$57; elsewhere US$160. Subscription renewals are subject to late fees. See www.ams.org/help-faq for more journal subscription information. Each number may be ordered separately; *please specify number* when ordering an individual number.

Back number information. For back issues see www.ams.org/bookstore.

Subscriptions and orders should be addressed to the American Mathematical Society, P. O. Box 845904, Boston, MA 02284-5904 USA. *All orders must be accompanied by payment.* Other correspondence should be addressed to 201 Charles Street, Providence, RI 02904-2294 USA.

Copying and reprinting. Individual readers of this publication, and nonprofit libraries acting for them, are permitted to make fair use of the material, such as to copy a chapter for use in teaching or research. Permission is granted to quote brief passages from this publication in reviews, provided the customary acknowledgment of the source is given.

Republication, systematic copying, or multiple reproduction of any material in this publication is permitted only under license from the American Mathematical Society. Requests for such permission should be addressed to the Acquisitions Department, American Mathematical Society, 201 Charles Street, Providence, Rhode Island 02904-2294 USA. Requests can also be made by e-mail to reprint-permission@ams.org.

Memoirs of the American Mathematical Society (ISSN 0065-9266) is published bimonthly (each volume consisting usually of more than one number) by the American Mathematical Society at 201 Charles Street, Providence, RI 02904-2294 USA. Periodicals postage paid at Providence, RI. Postmaster: Send address changes to Memoirs, American Mathematical Society, 201 Charles Street, Providence, RI 02904-2294 USA.

© 2010 by the American Mathematical Society. All rights reserved.
Copyright of individual articles may revert to the public domain 28 years
after publication. Contact the AMS for copyright status of individual articles.
This publication is indexed in *Science Citation Index*®, *SciSearch*®, *Research Alert*®,
CompuMath Citation Index®, *Current Contents*®/*Physical, Chemical & Earth Sciences*.
Printed in the United States of America.

∞ The paper used in this book is acid-free and falls within the guidelines
established to ensure permanence and durability.
Visit the AMS home page at http://www.ams.org/

10 9 8 7 6 5 4 3 2 1   15 14 13 12 11 10

Contents

Chapter 1.	Introduction	1
Chapter 2.	Erdős space and almost zero-dimensionality	5
Chapter 3.	Trees and $\mathbb{R}$-trees	7
Chapter 4.	Semi-continuous functions	11
Chapter 5.	Cohesion	21
Chapter 6.	Unknotting Lelek functions	25
Chapter 7.	Extrinsic characterizations of Erdős space	31
Chapter 8.	Intrinsic characterizations of Erdős space	39
Chapter 9.	Factoring Erdős space	49
Chapter 10.	Groups of homeomorphisms	51
Bibliography		61

Abstract

Let M be either a topological manifold, a Hilbert cube manifold, or a Menger manifold and let D be an arbitrary countable dense subset of M. Consider the topological group $\mathcal{H}(M,D)$ which consists of all autohomeomorphisms of M that map D onto itself equipped with the compact-open topology. We present a complete solution to the topological classification problem for $\mathcal{H}(M,D)$ as follows. If M is a one-dimensional topological manifold, then we proved in an earlier paper that $\mathcal{H}(M,D)$ is homeomorphic to $\mathbb{Q}^\omega$, the countable power of the space of rational numbers. In all other cases we find in this paper that $\mathcal{H}(M,D)$ is homeomorphic to the famed Erdős space $\mathfrak{E}$, which consists of the vectors in Hilbert space ℓ^2 with rational coordinates. We obtain the second result by developing topological characterizations of Erdős space.

Received by the editor October 21, 2005, and, in revised form, February 12, 2008.
Article electronically published on June 1, 2010; S 0065-9266(10)00579-X.
2000 *Mathematics Subject Classification*. Primary 57S05, 54F65.
Key words and phrases. Erdős space, almost zero-dimensional space, homeomorphism group, Lelek fan, upper semi-continuous function, cohesive space, $\mathbb{R}$-tree.

©2010 American Mathematical Society

CHAPTER 1

Introduction

All spaces under discussion are separable and metrizable. The main results of this paper were announced in Dijkstra and van Mill [20].

If X is compact then the standard topology on the group of homeomorphisms $\mathcal{H}(X)$ of X is the so-called compact-open topology (which coincides with the topology of uniform convergence). For noncompact locally compact spaces we give $\mathcal{H}(X)$ the topology that this group inherits from $\mathcal{H}(\alpha X)$, where αX is the one-point compactification. In either case we have that $\mathcal{H}(X)$ a Polish topological group. If A is a subset of a space X then $\mathcal{H}(X, A)$ stands for the subgroup $\{h \in \mathcal{H}(X) : h(A) = A\}$ of $\mathcal{H}(X)$.

Brouwer [11] showed that $\mathbb{R}$ is countable dense homogeneous, that is, for all countable dense subsets A and B of $\mathbb{R}$ there is an $h \in \mathcal{H}(\mathbb{R})$ with $h(A) = B$. It is not difficult to prove that every $\mathbb{R}^n$ has this property. In view of Brouwer's result it is a natural idea to investigate the group $\mathcal{H}(\mathbb{R}^n, \mathbb{Q}^n)$. It was shown in Dijkstra and van Mill [21] that the group $\mathcal{H}(\mathbb{R}, \mathbb{Q})$ is homeomorphic to the zero-dimensional space $\mathbb{Q}^\omega$, the countable infinite product of copies of the rational numbers $\mathbb{Q}$. In contrast, we showed in [21] (see also [15]) that $\mathcal{H}(\mathbb{R}^n, \mathbb{Q}^n)$ for $n \geq 2$ contains a closed copy of the famed Erdős space $\mathfrak{E}$ which is known to be one-dimensional, see [29]. This result led us to consider the question whether $\mathcal{H}(\mathbb{R}^n, \mathbb{Q}^n)$ (for $n \geq 2$) is in fact homeomorphic to Erdős space. We prove here that it is. We show that if D is a countable dense subset of a locally compact space X, then $\mathcal{H}(X, D)$ is an Erdős space factor, which means that $\mathcal{H}(X, D) \times \mathfrak{E}$ is homeomorphic to $\mathfrak{E}$. Under rather mild extra conditions, the group $\mathcal{H}(X, D)$ is found to be homeomorphic to Erdős space. This is the case if X contains a nonempty open subset homeomorphic to $\mathbb{R}^n$ for $n \geq 2$, an open subset of the Hilbert cube $\mathcal{Q}$, or an open subset of some universal Menger continuum. As an application it follows that if M is an at least 2-dimensional manifold (with or without boundary) and D is a countable dense subset of M, then $\mathcal{H}(M, D)$ is homeomorphic to Erdős space. These results can be found in Chapter 10.

Homeomorphism groups of manifolds are very well studied. Let $\mathbb{I}$ denote the interval $[0, 1]$ and let $\mathcal{H}^\partial(\mathbb{I}^n)$ stand for the subgroup of $\mathcal{H}(\mathbb{I}^n)$ consisting of homeomorphisms that fix the boundary of the n-cube $\mathbb{I}^n$. Anderson [5] proved that $\mathcal{H}^\partial(\mathbb{I})$ is homeomorphic to the separable Hilbert space ℓ^2 (see [8, Proposition VI.8.1] or [32]). It was shown by Luke and Mason [35] that $\mathcal{H}^\partial(\mathbb{I}^2)$ is an absolute retract, which implies that $\mathcal{H}^\partial(\mathbb{I}^2) \approx \ell^2$ (apply for instance Dobrowolski and Toruńczyk [26]). For $n \geq 3$ it is open whether $\mathcal{H}^\partial(\mathbb{I}^n)$ is an absolute retract. This is one of the most interesting open problems in infinite-dimensional topology. For the Hilbert cube $\mathcal{Q}$, that is for $n = \infty$, the analogous problem was solved by Ferry [30] and Toruńczyk [42]. They proved that $\mathcal{H}(\mathcal{Q})$ is homeomorphic to ℓ^2 (observe that $\mathcal{Q}$

has no boundary). For $2 < n < \infty$ it is unknown what the topological classification of $\mathcal{H}^\partial(\mathbb{I}^n)$ or $\mathcal{H}(\mathbb{I}^n)$ is. By our results, the subgroups $\mathcal{H}^\partial(\mathbb{I}^n, (\mathbb{Q} \cap \mathbb{I})^n)$ and $\mathcal{H}(\mathbb{I}^n, (\mathbb{Q} \cap \mathbb{I})^n)$ *are* known; they are homeomorphic to Erdős space.

Recall that *Erdős space* $\mathfrak{E}$ is the 'rational' Hilbert space, that is, the set of vectors in ℓ^2 the coordinates of which are all rational. This space was introduced by Hurewicz who asked to compute its dimension. Erdős [29] proved that $\mathfrak{E}$ is one-dimensional by establishing that every nonempty clopen subset of $\mathfrak{E}$ has diameter at least 1. This result, in combination with the obvious fact that $\mathfrak{E}$ is homeomorphic to $\mathfrak{E} \times \mathfrak{E}$, lends the space its importance in dimension theory. The paper [29] also features a closed subspace of ℓ^2 which has the same properties as we just listed for $\mathfrak{E}$. This space is now known as *complete Erdős space* $\mathfrak{E}_c$. The space $\mathfrak{E}_c$ surfaced in topological dynamics as the 'endpoint' set of several interesting objects; see Kawamura, Oversteegen, and Tymchatyn [31] for more information.

The heart of the present paper is formed by the Chapters 7 and 8 where we prove a series of increasingly powerful topological characterizations of Erdős space. Chapters 3–6 contain the lemmas that prepare the ground for the proofs of the characterization theorems. Chapters 9 and 10 contain the main applications of our characterization theorems. In Chapter 2 we introduce Erdős space with its basic properties. What sets Erdős space apart from familiar spaces is that in addition to the one-dimensional topology that it inherits from ℓ^2, an important role is played by the zero-dimensional topology that $\mathfrak{E}$ inherits from the product space $\mathbb{Q}^\omega$. This bitopological aspect was captured by Oversteegen and Tymchatyn [38] by the introduction of the class of almost zero-dimensional spaces of which $\mathfrak{E}$ and $\mathfrak{E}_c$ are universal elements. Chapter 3 contains information about $\mathbb{R}$-trees, the relevance of which to the Erdős spaces was established in [38] and developed further in [31]. Chapter 4 is devoted to semi-continuous functions and in particular to Lelek functions. The standard example of a Lelek function is an arclength function for a Lelek fan [33]. These functions are central to the understanding and characterization of Erdős spaces because of the proof by Kawamura, Oversteegen, and Tymchatyn [31] that complete Erdős space is homeomorphic to the endpoint set of the Lelek fan. In Chapter 5 we develop the cohesion concept. The idea of a cohesive space by which we mean a space that has a open cover consisting of sets that do not contain nonempty clopen subsets of the space is implicitly present in Erdős [29]. It is a very weak form of connectedness that plays an important role in the characterization theorems. In Chapter 6 we prove an "unknotting theorem" for Lelek functions. This theorem is made possible by the uniqueness of the Lelek fan as proved by Charatonik [14] and Bula and Oversteegen [12]. In Chapter 7 we present extrinsic characterizations of Erdős space, by which we mean characterizations that depend on particular imbeddings of the space in a space with more structure, in our case the graph of a Lelek function. We obtained inspiration for these characterizations from Sierpiński's [40] characterization of absolute $F_{\sigma\delta}$-spaces and from van Engelen's [28] characterization of the space $\mathbb{Q}^\omega$. From there we proceed by finding the more powerful intrinsic characterizations in Chapter 8, namely characterizations in terms of purely topological concepts that are internal to the space. In Chapter 9 we use the theorems in Chapter 8 to characterize the class of Erdős space factors. As a corollary we find that Erdős space is homeomorphic to its countable infinite power. Here we have a striking contrast with $\mathfrak{E}_c$ which is *not* homeomorphic to $\mathfrak{E}_c^\omega$, as was proved by Dijkstra, van Mill, and Steprāns [23]. We also find that Erdős space is homeomorphic to $\mathfrak{E}_c \times \mathbb{Q}^\omega$. Our main applications can be found in Chapter 10 where we demonstrate the power of our characterizations by deriving from them the above results on homeomorphism groups $\mathcal{H}(M, D)$ with relative ease.

We conclude with the observation that Erdős space started its career as a curious example in dimensional theory. It turns out however that it is a fundamental object that surfaces in many places. In addition, it allows for a useful and easily applied topological characterization just as several other fundamental objects in topology: the Cantor set (Brouwer [**10**]), the Hilbert cube (Toruńczyk [**43**]), Hilbert space (Toruńczyk [**44**]), and the universal Menger continua (Bestvina [**9**]).

CHAPTER 2

Erdős space and almost zero-dimensionality

Let $\mathbb{R}^+ = [0, \infty)$, $\mathbb{I} = [0, 1]$, and let $\omega = \{0\} \cup \mathbb{N}$. By a zero-dimensional space we mean a space with dim $= 0$ so in particular a non-empty space.

Let $p \in [1, \infty)$ and consider the Banach space ℓ^p. This space consists of all sequences $z = (z_0, z_1, z_2, \dots) \in \mathbb{R}^\omega$ such that $\sum_{i=0}^\infty |z_i|^p < \infty$. The topology on ℓ^p is generated by the norm $\|z\| = \left(\sum_{i=0}^\infty |z_i|^p\right)^{1/p}$. It is well known that this topology is the weakest topology that makes all the coordinate projections $z \mapsto z_i$ and the norm function continuous. This fact can also be formulated as follows: the norm topology on ℓ^p is generated by the product topology (that is inherited from $\mathbb{R}^\omega$) together with the sets $\{z \in \ell^p : \|z\| < t\}$ for $t > 0$. We extend the p-norm over $\mathbb{R}^\omega$ by putting $\|z\| = \infty$ when $z \in \mathbb{R}^\omega \setminus \ell^p$. Note that the norm as a function from $\mathbb{R}^\omega$ to $[0, \infty]$ is lower semi-continuous but not continuous.

DEFINITION 2.1. Let X be a space. A function $f: X \to [-\infty, \infty]$ is called *lower semi-continuous* (abbreviated LSC) if $f^{-1}((t, \infty])$ is open in X for every $t \in \mathbb{R}$. Similarly, f is called *upper semi-continuous* (abbreviated USC) if $f^{-1}([-\infty, t))$ is open in X for every $t \in \mathbb{R}$.

The following two spaces are featured in Erdős [**29**]: *Erdős space*

(2.1) $$\mathfrak{E} = \{x \in \ell^2 : x_i \in \mathbb{Q} \text{ for each } i \in \omega\}$$

and *complete Erdős space*

(2.2) $$\mathfrak{E}_c = \{x \in \ell^2 : x_i \in \{0\} \cup \{1/n : n \in \mathbb{N}\} \text{ for each } i \in \omega\}.$$

The name complete Erdős space was introduced by Kawamura, Oversteegen, and Tymchatyn [**31**] who used the following representation of that space:

(2.3) $$\mathfrak{E}'_c = \{x \in \ell^2 : x_i \notin \mathbb{Q} \text{ for each } i \in \omega\}.$$

Note that $\mathfrak{E}_c$ is closed in ℓ^2 and that $\mathfrak{E}'_c$ is not closed but a G_δ-set in ℓ^2, thus both spaces are topologically complete. It is a consequence of the characterization theorem for complete Erdős space that is presented in [**31**] that $\mathfrak{E}_c$ and $\mathfrak{E}'_c$ are homeomorphic; see [**16**]. The properties of complete Erdős space have been studied extensively in [**31**], [**25**], [**23**], and [**22**].

Let $\mathcal{T}$ stand for the zero-dimensional topology that $\mathfrak{E}$ inherits from $\mathbb{Q}^\omega$. Observe that $\mathcal{T}$ is weaker than the norm topology and hence that $\mathfrak{E}$ is totally disconnected. We have by the remark above that the graph of the norm function, when seen as a function from $(\mathfrak{E}, \mathcal{T})$ to $\mathbb{R}^+$, is homeomorphic to $\mathfrak{E}$. So, informally, we can think of $\mathfrak{E}$ as a 'zero-dimensional space with some LSC function declared continuous'. We find it convenient to work with USC rather than LSC functions and we therefore define $\eta : \mathbb{Q}^\omega \to \mathbb{R}^+$ by

(2.4) $$\eta(z) = 1/(1 + \|z\|),$$

5

where $1/\infty = 0$.

There is an interesting connection between the two topologies on $\mathfrak{E}$ that we would like to draw attention to. Because the norm is LSC on $\mathbb{R}^\omega$ every closed ε-ball in $\mathfrak{E}$ is also closed in the zero-dimensional space $\mathbb{Q}^\omega$. Thus we have that every point in $\mathfrak{E}$ has arbitrarily small neighbourhoods which are intersections of clopen sets.

DEFINITION 2.2. A subset A of a space X is called a *C-set in X* if A can be written as an intersection of clopen subsets of X. A space is called *almost zero-dimensional* if every point of the space has a neighbourhood basis consisting of C-sets of the space. If Z is a set that contains X then we say that a (separable metric) topology $\mathcal{T}$ on Z *witnesses the almost zero-dimensionality of* X if $\dim(Z,\mathcal{T}) \leq 0$, $O \cap X$ is open in X for each $O \in \mathcal{T}$, and every point of X has a neighbourhood basis in X consisting of sets that are closed in $(Z,\mathcal{T})$. We will also say that the space $(Z,\mathcal{T})$ is a witness to the almost zero-dimensionality of X.

Thus $\mathfrak{E}$ is almost zero-dimensional. In fact, it is a universal object for the class of almost zero-dimensional spaces, see Theorem 4.15. The space $\mathbb{Q}^\omega$ is a witness to the almost zero-dimensionality of Erdős space. More generally, if $\varphi : Z \to \mathbb{R}$ is a USC or LSC function with a zero-dimensional domain, then Z is easily seen to be a witness to the almost zero-dimensionality of the graph of φ.

REMARK 2.3. Observe that every C-set is closed and that finite unions and finite intersections of C-sets are also C-sets. The concept of an almost zero-dimensional space is due to Oversteegen and Tymchatyn [38]. The definition given here is easier to use than the original one in [38] and shown to be equivalent in Dijkstra, van Mill, and Steprāns [23]. Note that almost zero-dimensionality is hereditary. It is proved in [38] that every almost zero-dimensional space is at most one-dimensional; see also Levin and Pol [34] and Abry and Dijkstra [1].

Lemma 2.4 of the preprint version is now Lemma 4.10.

REMARK 2.4. A space X is almost zero-dimensional if and only if there is a topology on X witnessing this fact. This fact can easily be seen as follows. If X is almost zero-dimensional then there exists a collection $\mathcal{B}$ of subsets of X such that

(1) for every $x \in X$ and every neighbourhood U of x there is a $B \in \mathcal{B}$ such that $x \in \operatorname{int} B \subset B \subset U$ and
(2) every $B \in \mathcal{B}$ is an intersection of clopen subsets of X.

Since X is separable metric we may assume that $\mathcal{B}$ is countable and we can find for each $B \in \mathcal{B}$ a countable collection $\mathcal{C}_B$ of clopen sets such that $B = \bigcap \mathcal{C}_B$. Then it is easily verified that $\{C, X \setminus C : C \in \mathcal{C}_B, B \in \mathcal{B}\}$ is a subbasis for a separable metric topology that witnesses the almost zero-dimensionality of X.

REMARK 2.5. Let $(X, \mathcal{T})$ be a witness to the almost zero-dimensionality of some space X and let O be open in X. Since X is separable metric we can write O as a union of countably many sets that are closed in the topology $\mathcal{T}$. Thus every open set of X is F_σ in the witness topology $\mathcal{T}$ and dually every closed set is G_δ with respect to $\mathcal{T}$.

CHAPTER 3

Trees and $\mathbb{R}$-trees

An $\mathbb{R}$-*tree* is a locally arcwise connected and uniquely arcwise connected space. Let X be a uniquely arcwise connected space. If $x, y \in X$ with $x \neq y$ then $[x, y]$ denotes the unique arc in X that has x and y as endpoints; $[x, x]$ denotes the singleton $\{x\}$. We shall also use $[x, y) = [x, y] \setminus \{y\}$, and $(x, y) = [x, y) \setminus \{x\}$. We define the set of *interior* points of X by $\mathfrak{i}X = \{z \in X : z \in (x, y) \text{ for some } x, y \in X\}$. The set of *endpoints* of X is $\mathfrak{e}X = X \setminus \mathfrak{i}X$. If $p, x, y \in X$ then there is a unique $z \in X$ such that $[p, x] \cap [p, y] = [p, z]$. We define the *meet* function $\wedge_p \colon X \times X \to X$ by setting $x \wedge_p y = z$. Note that $[x, z] \cup [z, y] = [x, y]$ so $x \wedge_p y = p \wedge_y x = y \wedge_x p$.

Mayer and Oversteegen [36] proved that any $\mathbb{R}$-tree $\mathbb{T}$ admits a *convex* metric ρ, that is a metric that generates the topology of $\mathbb{T}$ and that has the property $\rho(x, y) + \rho(y, z) = \rho(x, z)$ whenever $y \in [x, z] \subset \mathbb{T}$.

Let $\mathbb{T}$ be an $\mathbb{R}$-tree. The *weak* topology on $\mathbb{T}$ is the topology that is generated by the following subbasis

(3.1) $\mathcal{S} = \{C \colon C \text{ is a component of } \mathbb{T} \setminus \{x\} \text{ for some } x \in \mathbb{T}\}.$

Note that since $\mathbb{T}$ is locally arcwise connected every $C \in \mathcal{S}$ is an open subset of $\mathbb{T}$ that is arcwise connected and therefore itself an $\mathbb{R}$-tree. So the weak topology is weaker, but not necessarily strictly weaker, than the original topology. Let $\mathbb{T}_w$ stand for the set $\mathbb{T}$ equipped with the weak topology and let $\mathfrak{e}\mathbb{T}_w$ denote the set $\mathfrak{e}\mathbb{T}$ with the topology that is inherited from $\mathbb{T}_w$. Note that it follows from the next lemma that on compact subsets of $\mathbb{T}$ (such as arcs) the weak topology coincides with the given topology.

LEMMA 3.1. *If $\mathbb{T}$ is an $\mathbb{R}$-tree then $\mathbb{T}_w$ is a separable metric space and $\dim(\mathfrak{e}\mathbb{T}_w) \leq 0$.*

PROOF. Let ρ be a convex metric on $\mathbb{T}$. By the definition of the subbasis $\mathcal{S}$ the space $\mathbb{T}_w$ is obviously T_1. Consider a countable dense subset D of $\mathbb{T}$ and select for every $x, y \in D$ with $x \neq y$ a countable dense subset Q_{xy} in (x, y). Define the countable set

(3.2) $Q = \bigcup \{Q_{xy} \colon x, y \in D \text{ with } x \neq y\}$

and the following subcollection of $\mathcal{S}$

(3.3) $\mathcal{S}' = \{C \colon C \text{ is a component of } \mathbb{T} \setminus \{x\} \text{ for some } x \in Q\}.$

Since the components of $\mathbb{T} \setminus \{x\}$ form a pairwise disjoint open collection in the separable space $\mathbb{T}$ we have that this collection is countable and hence $\mathcal{S}'$ is countable as well. Let $p, x \in \mathbb{T}$ and let C be a component of $\mathbb{T} \setminus \{x\}$ such that $p \in C$. Note that $[p, x) \subset C$. Select $q, y \in D$ such that $\rho(p, q), \rho(x, y) < \rho(p, x)/2$. Put $z = x \wedge_p y$ and $r = p \wedge_x q$. Since $\rho(x, z) + \rho(z, y) = \rho(x, y)$ we have $\rho(x, z) < \rho(p, x)/2$. Analogously, we find that $\rho(p, r) < \rho(p, x)/2$. Consequently, $[q, y] \cap [p, x] = [r, z]$

is a nondegenerate arc. So $(r,z) \subset (p,x)$ and we can find an $a \in Q_{qy} \subset Q$ that is contained in (p,x). Consider the component K of $\mathbb{T} \setminus \{a\}$ that contains p. If $b \in K$ then $[p,b] \subset K$ and hence $[p,b] \cap [p,x]$ is a (possibly degenerate) subarc of $[p,x]$ that contains p but not a. So $[p,b]$ cannot contain x and hence $b \in C$. Consequently, $\overline{K} = K \cup \{a\} \subset C$. So have proved that $\mathbb{T}_w$ is regular and has a countable subbasis $\mathcal{S}'$, in short, $\mathbb{T}_w$ is a separable metric space.

Since every element x of Q is an interior point of $\mathbb{T}$ we have that $\{C \cap \mathfrak{e}\mathbb{T} : C$ a component of $\mathbb{T} \setminus \{x\}\}$ forms a clopen partition of $\mathfrak{e}\mathbb{T}$. So every $C \cap \mathfrak{e}\mathbb{T}$ where $C \in \mathcal{S}'$ is clopen in $\mathfrak{e}\mathbb{T}_w$. Consequently, $\mathfrak{e}\mathbb{T}_w$ is zero-dimensional. □

The following lemma refines the result that $\mathfrak{e}\mathbb{T}$ is almost zero-dimensional, which was proved by Oversteegen and Tymchatyn [**38**].

LEMMA 3.2. *Let $(\mathbb{T}, \rho)$ be an $\mathbb{R}$-tree with a convex metric. Let $p \in \mathbb{T}$ be a fixed point and let $\varphi \colon \mathbb{T}_w \to \mathbb{R}^+$ be defined by $\varphi(x) = \rho(p,x)$. Then φ is an LSC function such that the natural projection π from the graph of φ to $\mathbb{T}$ is a homeomorphism. Consequently, $\mathfrak{e}\mathbb{T}_w$ witnesses the almost zero-dimensionality of $\mathfrak{e}\mathbb{T}$.*

PROOF. First we show that φ is LSC. Let $x \in \mathbb{T}$ and $t \in \mathbb{R}^+$ such that $\varphi(x) > t$. Then $x \neq p$ and we can select a $y \in (p,x)$ such that $\varphi(y) > t$. Let C be the component of $\mathbb{T} \setminus \{y\}$ that contains x and let $a \in C$. Since $[a,x] \subset C$ we have $a \wedge_p x \in (y,x]$ so $\varphi(a) \geq \varphi(a \wedge_p x) > \varphi(y) > t$.

Note that the graph of φ, $G = \{(x, \varphi(x)) \colon x \in \mathbb{T}_w\}$, inherits the topology from $\mathbb{T}_w \times \mathbb{R}^1$ and that $\pi(x, \varphi(x)) = x$. Since $\rho(p,x)$ is continuous as a function on $\mathbb{T}$ and the topology on $\mathbb{T}_w$ is weaker than the topology on $\mathbb{T}$ we have that π^{-1} is continuous. To show that π is continuous let $x \in \mathbb{T}_w$ and let $\varepsilon > 0$. It is obvious that π is continuous at $(p,0)$ so assume that $x \neq p$. Select a $y \in (p,x)$ such that $\rho(x,y) \leq \varepsilon/3$ and let C be the component of $\mathbb{T} \setminus \{y\}$ that contains x. Note that

(3.4) $\qquad U = \{(a, \varphi(a)) \colon a \in C \text{ and } \varphi(a) < \varphi(x) + \varepsilon/3\}$

is an open neighbourhood of $(x, \varphi(x))$ in G. Let $(z, \varphi(z)) \in U$ and put $a = x \wedge_p z$. We have $\rho(p,x) = \rho(p,a) + \rho(a,x)$ and $\rho(p,z) = \rho(p,a) + \rho(a,z)$ thus $\varphi(z) - \varphi(x) = \rho(p,z) - \rho(p,x) = \rho(a,z) - \rho(a,x) < \varepsilon/3$. So we conclude that

(3.5) $\qquad \rho(a,z) < \rho(a,x) + \varepsilon/3.$

Since $[x,z] = [x,a] \cup [a,z] \subset C$ we have that $a \in (y,x]$ and so by (3.5),

(3.6) $\qquad \rho(x,z) = \rho(x,a) + \rho(a,z) < 2\rho(x,a) + \varepsilon/3 < 2\rho(x,y) + \varepsilon/3 \leq \varepsilon.$

Finally, we note since φ is LSC closed ε-balls around any p are closed with respect to the weak topology so using also Lemma 3.1 we find that $\mathfrak{e}\mathbb{T}_w$ witnesses the almost zero-dimensionality of $\mathfrak{e}\mathbb{T}$. □

DEFINITION 3.3. If A is a nonempty set then $A^{<\omega}$ denotes the set of all finite strings of elements of A, including the null string $\emptyset$. If $s \in A^{<\omega}$ then $|s|$ denotes its *length*. In this context the set A is called an *alphabet*. Let A^ω denote the set of all infinite strings of elements of A. If $s \in A^{<\omega}$ and $\sigma \in A^{<\omega} \cup A^\omega$, then we put $s \prec \sigma$ if s is an initial substring of σ, that is, there is a $\tau \in A^{<\omega} \cup A^\omega$ with $s^\frown \tau = \sigma$, where $\frown$ denotes concatenation of strings. If $\sigma \in A^{<\omega} \cup A^\omega$ and $k \in \omega$, then $\sigma \upharpoonright k \in A^{<\omega}$ is characterized by $\sigma \upharpoonright k \prec \sigma$ and $|\sigma \upharpoonright k| = k$.

DEFINITION 3.4. A *tree* T on an alphabet A is a subset of $A^{<\omega}$ that is closed under initial segments, i.e., if $s \in T$ and $t \prec s$ then $t \in T$. Elements of T are called *nodes*. An *infinite branch* of T is an element σ of A^ω such that $\sigma{\restriction}k \in T$ for every $k \in \omega$. The *body* of T, written as $[T]$ is the set of all infinite branches of T. If $s, t \in T$ are such that $s \prec t$ and $|t| = |s| + 1$, then we say that t is an *immediate successor* of s and $\operatorname{succ}(s)$ denotes the set of immediate successors of s in T. A tree is called *pruned* if $\operatorname{succ}(s) \neq \emptyset$ for each node s.

If S and T are trees over A respectively B, then we define the product tree $S * T$ as follows. If $s = a_1 \ldots a_l \in S$ and $t = b_1 \ldots b_l \in T$ are two strings of equal length, then we define the string $s * t$ over $A \times B$ by $s * t = (a_1, b_1) \ldots (a_l, b_l)$. We define $S * T = \{s * t : s \in S, t \in T, |s| = |t|\}$ and note that it is a tree over $A \times B$.

Let T be a tree over a countable set A. Consider the Banach space ℓ^1 with norm $\|z\| = \sum_{i=0}^{\infty} |z_i|$. Let $\{e^k : k \in \omega\}$ be the standard basis for ℓ^1 given by $e_i^k = 0$ if $k \neq i$ and $e_i^k = 1$ if $k = i$. If $z, z' \in \ell^1$ then we let $\langle z, z' \rangle$ denote the line segment $\{z + t(z' - z) : t \in \mathbb{I}\}$. Let $\nu : T \to \omega$ be an injection. We define a function $\gamma : T \cup [T] \to \ell^1$ as follows: if $\sigma \in T \cup [T]$ then

$$(3.7) \qquad \gamma(\sigma) = \sum_{k=1}^{|\sigma|} 2^{-k} e^{\nu(\sigma{\restriction}k)}.$$

We define

$$(3.8) \qquad T_\mathbb{R} = \gamma([T]) \cup \bigcup \{\langle \gamma(s), \gamma(t) \rangle : s \in T \text{ and } t \in \operatorname{succ}(s)\}.$$

We note the following fact: $\|\gamma(\sigma)\| = 1 - 2^{-|\sigma|}$ for $\sigma \in T \cup [T]$, where $2^{-\infty} = 0$. Also observe that if we equip $[T]$ with the topology that this set inherits from the product space A^ω with A discrete, then $\gamma{\restriction}[T]$ is an imbedding. It is easily verified that the norm produces a convex metric ρ on $T_\mathbb{R}$ and that this space is an $\mathbb{R}$-tree. Note also that the weak topology on $T_\mathbb{R}$ coincides with the norm topology. If the tree T is pruned and if $\operatorname{succ}(\emptyset)$ has at least two elements, then $eT_\mathbb{R} = \gamma([T]) = T_\mathbb{R} \cap \{z \in \ell^1 : \|z\| = 1\}$.

Oversteegen and Tymchatyn [38] have shown that a space is almost zero-dimensional if and only if it is homeomorphic to eT for some $\mathbb{R}$-tree T. We need a slightly more precise version of that result. Our refinement can be extracted from the proof in [38] but we give a direct proof here for the sake of completeness.

LEMMA 3.5. *If X is almost zero-dimensional as witnessed by a topology $\mathcal{T}$ on X, then there are an $\mathbb{R}$-tree $\mathbb{T}$ and a homeomorphism $h : X \to e\mathbb{T}$ such that also $h : (X, \mathcal{T}) \to e\mathbb{T}_w$ is a homeomorphism.*

PROOF. We may assume that X has at least two elements. Let $\{C_i : i \in \mathbb{N}\}$ be a basis for $\mathcal{T}$ consisting of clopen sets such that both C_1 and $X \setminus C_1$ are not empty. We put $A = \{0, 1\}$ and consider the tree $A^{<\omega}$. We define

$$(3.9) \qquad D_s = \bigcap \{C_i : i \leq l, s_i = 1\} \cap \bigcap \{X \setminus C_i : i \leq l, s_i = 0\},$$

where $s = s_1 \ldots s_l \in A^{<\omega}$ and $\bigcap \emptyset = X$. Let $T = \{s \in A^{<\omega} : D_s \neq \emptyset\}$ and note that T is a pruned tree and that $|\operatorname{succ}(\emptyset)| = 2$. Thus $eT_\mathbb{R} = \gamma([T])$. For every $x \in X$ we define $\tau(x) \in [T]$ by $x \in \bigcap_{k=0}^{\infty} D_{\tau(x){\restriction}k}$. Clearly, τ is an imbedding of $(X, \mathcal{T})$ into $[T]$ with the topology that is inherited from the Cantor set A^ω.

By the fact that $\mathcal{T}$ is a witness topology we can find a countable collection $\mathcal{B}$ of subsets of X such that

(a) for every $x \in X$ and every neighbourhood U of x there is a $B \in \mathcal{B}$ such that $x \in \operatorname{int} B \subset B \subset U$ and

(b) every $B \in \mathcal{B}$ is closed with respect to $\mathcal{T}$.

If $Y \subset \mathfrak{e}T_{\mathbb{R}}$, $s \in T$, and $\varepsilon > 0$, then we define the following subsets of $\mathfrak{i}T_{\mathbb{R}}$:

(3.10) $$\Delta(Y,s) = \bigcup \{(y, \gamma(s)]: y \in Y\}$$

and

(3.11) $$\Delta_\varepsilon(Y,s) = \{z \in T_{\mathbb{R}}: \rho(z, \Delta(Y,s)) < \varepsilon(1 - \|z\|)\}.$$

Note that $\Delta_\varepsilon(Y,s)$ is open in $T_{\mathbb{R}}$ and that $\Delta_\varepsilon^+(Y,s) = Y \cup \Delta_\varepsilon(Y,s)$ is an $\mathbb{R}$-tree for $\varepsilon \leq 1$ by the convexity of ρ.

We now consider $\mathbb{T} = \gamma(\tau(X)) \cup \mathfrak{i}T_{\mathbb{R}}$ and we let the topology on $\mathbb{T}$ be generated by the basis

(3.12) $\quad \{O \cap \mathbb{T}: O \text{ open in } T_{\mathbb{R}}\} \cup \{\Delta_{2^{-j}}^+(\gamma(\tau(\operatorname{int} B)), s): j \in \mathbb{N}, B \in \mathcal{B}, s \in T\}.$

Observe that $\mathbb{T}$ is a second countable Hausdorff space that is uniquely arcwise connected and locally arcwise connected. Clearly, we have that $\gamma \circ \tau: X \to \gamma(\tau(X)) = \mathfrak{e}\mathbb{T}$ is a homeomorphism. Since the weak topology is generated by points in the interior of an $\mathbb{R}$-tree we have that the weak topology on $\mathbb{T}$ coincides with the topology inherited from $T_{\mathbb{R}} = (T_{\mathbb{R}})_w$. Consequently, we have that $\gamma \circ \tau: (X, \mathcal{T}) \to \mathfrak{e}\mathbb{T}_w$ is also a homeomorphism.

The only thing left is to verify that $\mathbb{T}$ is a regular space which implies that it is separable metric. We obviously only have to consider points $\gamma(\tau(x)) = z \in \mathfrak{e}\mathbb{T}$ that are contained in basic sets of the form $U = \Delta_{2^{-j}}^+(\gamma(\tau(\operatorname{int} B)), s)$. Let $B' \in \mathcal{B}$ be such that $x \in \operatorname{int} B' \subset B' \subset \operatorname{int} B$ and consider the closed set

(3.13) $$V = \{z \in \mathbb{T}: \rho(z, \Delta(\gamma(\tau(B')), s)) \leq 2^{-j-1}(1 - \|z\|)\}.$$

Since V contains $\Delta_{2^{-j-1}}^+(\gamma(\tau(\operatorname{int} B')), s)$ it is a neighbourhood of x in $\mathbb{T}$. It now suffices to show that V is contained in U. Let $z \in V$. If $z' \in \mathfrak{i}\mathbb{T}$ then clearly $z' \in U$. Now let $z' \in \mathfrak{e}\mathbb{T}$. Then $\|z'\| = 1$ and $\rho(z', \Delta(\gamma(\tau(B')), s)) = 0$. Let $0 < \varepsilon < 1 - \|\gamma(s)\|)$ and select a $b \in B'$ such that $\rho(z', (\gamma(\tau(b)), \gamma(s)]) < \varepsilon$. Then by convexity of ρ we have $\rho(z', y) < \varepsilon$, where $y = \gamma(\tau(b)) \wedge_s z'$. Note that $\|y\| > 1 - \varepsilon > \|\gamma(s)\|$ and that $y = \gamma(s')$ for some $s' \in T$. We then have that $s' \prec \tau(b)$ and hence $\rho(y, \gamma(\tau(b))) = 2^{-|s'|} = 1 - \|y\| < \varepsilon$. So $\rho(z', \gamma(\tau(b))) < 2\varepsilon$ and we may conclude that $\rho(z', \gamma(\tau(B'))) = 0$. Since B' is closed in $(X, \mathcal{T})$ we have that $\gamma(\tau(B'))$ is closed in $(\mathfrak{e}\mathbb{T}, \rho)$ and hence $z' \in \gamma(\tau(B')) \subset U$. □

CHAPTER 4

Semi-continuous functions

DEFINITION 4.1. Let $\varphi, \psi\colon X \to \mathbb{R}^+$ be such that X is a space and $\psi(x) \leq \varphi(x)$ for all $x \in X$. We define

(4.1) $$G_\psi^\varphi = \{(x, \varphi(x))\colon x \in X \text{ and } \varphi(x) > \psi(x)\}$$

and

(4.2) $$L_\psi^\varphi = \{(x, t)\colon x \in X \text{ and } \psi(x) \leq t \leq \varphi(x)\}$$

both equipped with the topology inherited from $X \times \mathbb{R}^+$. We say that φ is a *Lelek function with bias* ψ if X is zero-dimensional, φ and ψ are USC, $X' = \{x \in X\colon \psi(x) < \varphi(x)\}$ is dense in X, and G_ψ^φ is dense in $L_{\psi\restriction X'}^{\varphi\restriction X'}$. If φ is a Lelek function with bias 0, then φ is simply called a *Lelek function*.

For a space X we let $\pi_1\colon X \times \mathbb{R}^+ \to X$ and $\pi_2\colon X \times \mathbb{R}^+ \to \mathbb{R}^+$ denote the projections.

REMARK 4.2. Let $\varphi\colon X \to \mathbb{R}^+$ be a USC function with $\dim X = 0$. Note that φ is continuous at points from $\varphi^{-1}(0)$. Let Y be the graph of φ with the topology that is lifted from X (so Y and X are homeomorphic). Let $(x, \varphi(x)) \in G_0^\varphi$ and note that since φ is USC a basic neighbourhood of the point has the form $B(U, t) = G_0^\varphi \cap (U \times [t, \infty))$, where U is a clopen neighbourhood of x in X and $0 < t < \varphi(x)$. Note that

(4.3) $$\pi_1(B(U, t)) = U \cap \varphi^{-1}([t, \infty))$$

is a closed subset of X so $B(U, t)$ is closed in Y. This makes Y a witness to the almost zero-dimensionality of G_0^φ. So every open subset of G_0^φ is an F_σ-set in Y.

REMARK 4.3. If φ is a Lelek function and O is an open subset of G_0^φ, then $\pi_1(O)$ is of the first category in itself. This result can be seen as follows. We can cover O with countably many sets of the form $B(U, t)$ such that there is an $s < t$ with $B(U, s) \subset O$. Let $(x, \varphi(x))$ be an element of such a $B(U, t)$ and let V be a neighbourhood of x in X. Since φ is Lelek there must be a $y \in U \cap V$ such that $s < \varphi(y) < t$ and hence $y \in \pi_1(O) \setminus \pi_1(B(U, t))$ by (4.3). So $\pi_1(B(U, t))$ is nowhere dense in $\pi_1(O)$ and $\pi_1(O)$ is of the first category in itself.

PROPOSITION 4.4. *If $\varphi\colon X \to \mathbb{R}^+$ is a Lelek function with X topologically complete, then every nonempty clopen subset C of G_0^φ fails to be closed in the full graph of φ and the projection $\pi_2(C)$ is an interval that has 0 as one of its endpoints.*

PROOF. Put $Z = X \setminus \pi_1(G_0^\varphi \setminus C)$ and note that by Remark 4.2 the space Z is a G_δ-set in X and hence topologically complete. Also, $\pi_1(C)$ is a first category F_σ-subset of Z by Remarks 4.2 and 4.3. Thus $\pi_1(C)$ is not closed in Z and we can find an $x \in Z \cap (\overline{\pi_1(C)} \setminus \pi_1(C))$. Since $Z \setminus \pi_1(C) = \varphi^{-1}(0)$ we have $\varphi(x) = 0$. Recalling that φ is continuous at x we find that $(x, 0)$ is a cluster point of C.

Now let $(x, \varphi(x)) \in C$ be such that there is a $t \notin \pi_2(C)$ with $0 < t < \varphi(x)$. Then $C \cap (X \times [t, \infty)) = C \cap (X \times (t, \infty))$ is a clopen, nonempty subset of G_0^φ that is closed in the graph of φ, a contradiction. □

DEFINITION 4.5. If $\varphi \colon X \to \mathbb{R}$ then we define
(4.4) $$M(\varphi) = \sup\{|\varphi(x)| : x \in X\} \in [0, \infty].$$
If $X = \emptyset$ then we use the convention $M(\varphi) = \sup \emptyset = 0$.

REMARK 4.6. A USC function $\varphi \colon X \to \mathbb{R}^+$ with $\dim X = 0$ is a Lelek function if and only if G_0^φ is dense in L_0^φ. For the "if" part note that $X' = \pi_1(G_0^\varphi)$ and $X = \pi_1(L_0^\varphi)$. For the "only if" part use the fact that φ is continuous in points of $\varphi^{-1}(0) = X \setminus X'$.

We obviously have that the domain of a Lelek function φ is dense in itself and that $M(\varphi) > 0$. Lelek functions with compact domain C exist (see Lelek [33]). The domain C must be a Cantor set and $\varphi(C) = [0, M(f)]$. If φ is a Lelek function with a compactum C as domain and we identify the set $C \times \{0\}$ to a point in L_0^φ, then we obtain a *Lelek fan*. Note that G_0^φ is the endpoint set of the fan. According to Kawamura, Oversteegen and Tymchatyn [31] we have in this case that G_0^φ is homeomorphic to complete Erdős space.

DEFINITION 4.7. Let $\varphi \colon X \to [0, \infty]$ be a function and let X be a subset of a metric space (Y, d). We define $\text{ext}_Y \varphi \colon Y \to [0, \infty]$ by
(4.5) $$(\text{ext}_Y \varphi)(y) = \lim_{\varepsilon \searrow 0} M(\varphi{\restriction}(X \cap U_\varepsilon(y))) \quad \text{for } y \in Y,$$
where $U_\varepsilon(y) = \{x \in Y : d(x, y) < \varepsilon\}$. Observe that $M(\varphi{\restriction}(X \cap U_\varepsilon(y)))$ is a nondecreasing function of ε so the limit is always well-defined. Note that the metric on Y is mentioned strictly for the sake of convenience and that the definition of $\text{ext}_Y \varphi$ does not depend on the choice of d. Note also that the image of $\text{ext}_Y \varphi$ is contained in the closure of $\varphi(X) \cup \{0\}$.

LEMMA 4.8. *Let X be a subset of a space Y and let $\varphi \colon X \to [0, \infty]$ be a function. Put $\psi = \text{ext}_Y \varphi$.*

(a) *Then ψ is a USC function and $\psi{\restriction}Y \setminus \overline{X} = 0$. If φ is USC then ψ extends φ and the graph of φ is dense in the graph of $\psi{\restriction}\overline{X}$.*

(b) *If φ is a bounded Lelek function, $\dim Y = 0$, and X is dense in Y, then G_0^φ is dense in L_0^ψ (thus ψ is also a Lelek function).*

(c) *If $\dim X = 0$ and $\mathcal{F}$ is a countable collection of closed subsets of X, then there is a zero-dimensional compactification C of X such that $(\text{ext}_C \varphi){\restriction}\overline{F} = \text{ext}_{\overline{F}}(\varphi{\restriction}F)$ for each $F \in \mathcal{F}$, where $\overline{F}$ is the closure in C.*

PROOF. It is clear that the closure of φ in $X \times [0, \infty]$ contains $\psi{\restriction}\overline{X}$ and that $\psi{\restriction}Y \setminus \overline{X} = 0$. In order to prove that ψ is USC let $y \in Y$ and let $t > \psi(y)$. Then there is an $\varepsilon > 0$ such that $s = M(\varphi{\restriction}(X \cap U_\varepsilon(y))) < t$. Obviously, every element $y' \in U_\varepsilon(y)$ has the property $\psi(y') \leq s < t$. Now let φ be USC. If $y \in X$ and $\varepsilon > 0$, then since φ is USC there is a $\delta > 0$ such that $\varphi(x) < \varphi(y) + \varepsilon$ for every $x \in X \cap U_\delta(y)$. Thus $\varphi(y) \leq M(\varphi{\restriction}(X \cap U_\delta(y))) \leq \varphi(y) + \varepsilon$ and hence $\psi(y) = \varphi(y)$.

For point (b) let φ be a bounded Lelek function and hence $\psi(Y) \subset \mathbb{R}^+$. Let $(y, t) \in L_0^\psi$ and let $\varepsilon > 0$. Note that $M(\varphi{\restriction}(X \cap U_\varepsilon(y))) \geq \psi(y) \geq t$ so we can find an $x \in X \cap U_\varepsilon(y)$ with $\varphi(x) > t - \varepsilon/2$, where we used the assumption that X is dense in Y. Put $s = \max\{0, t - \varepsilon/2\}$ thus $0 \leq s \leq \varphi(x)$ and $|t - s| \leq \varepsilon/2$. Since φ is Lelek there is a $z \in X \cap U_\varepsilon(y)$ such that $\varphi(z) > 0$ and $|s - \varphi(z)| < \varepsilon/2$. So $(z, \varphi(z))$ is a point in G_0^φ that is ε-close to (y, t) with respect to the max metric on $Y \times \hat{\mathbb{R}}$.

For point (c) consider a closed collection $\{F_i : i \in \mathbb{N}\}$ in X. We will construct recursively a sequence $\mathcal{B}_1 \subset \mathcal{B}_2 \subset \cdots$ of countable boolean algebras consisting of clopen subsets of X. Let $\mathcal{B}_1$ be the boolean algebra that is generated by some countable clopen basis for the topology on X. Assume that $\mathcal{B}_n$ has been constructed. Let $B \in \mathcal{B}_n$ and let $i \in \mathbb{N}$. If $B \cap F_i = \emptyset$ then we put $C(B, i) = B$. If $B \cap F_i \neq \emptyset$ then we consider the open neighbourhood

$$(4.6) \qquad U = \{x \in X : \varphi(x) < 2^{-n} + M(\varphi{\restriction}(B \cap F_i))\}$$

of the closed set $B \cap F_i$. We now let $C(B, i)$ be a clopen neighbourhood of $B \cap F_i$ that is contained in U. We let $\mathcal{B}_{n+1}$ be the boolean algebra that is generated by $\mathcal{B}_n \cup \{C(B, i) : B \in \mathcal{B}_n, i \in \mathbb{N}\}$.

Let C be the Stone space that corresponds to the boolean algebra $\mathcal{B} = \bigcup_{n=1}^\infty \mathcal{B}_n$. Note that C is a metrizable and zero-dimensional compactification of X. Let $i \in \mathbb{N}$ and let $y \in \overline{F_i} \setminus F$. It is clear that $(\text{ext}_{\overline{F_i}}(\varphi{\restriction}F_i))(y) \leq (\text{ext}_C \varphi)(y)$. Let V be a neighbourhood of y in C and let $k \in \mathbb{N}$. We may assume that V is a basic neighbourhood so that $V \cap X = B$ for some $B \in \mathcal{B}$. Let $n \geq k$ such that $B \in \mathcal{B}_n$. Since $y \in \overline{F_i}$ we have $B \cap F_i \neq \emptyset$ and we consider $C(B, i) \in \mathcal{B}_{n+1}$. Since $C(B, i)$ is an element of $\mathcal{B}$ that contains $B \cap F_i$ we have that $\overline{C(B, i)}$ is a neighbourhood of y. Consequently,

$$(4.7) \quad (\text{ext}_C \varphi)(y) \leq M(\varphi{\restriction}C(B, i)) \leq 2^{-n} + M(\varphi{\restriction}(B \cap F_i))$$
$$\leq 2^{-k} + M(\varphi{\restriction}(V \cap F_i)).$$

Since V and k are arbitrary we have that $(\text{ext}_C \varphi)(y) \leq (\text{ext}_{\overline{F_i}}(\varphi{\restriction}F_i))(y)$. $\square$

LEMMA 4.9. *If $\varphi, \psi \colon X \to \mathbb{R}^+$ are USC functions such that $\psi(x) \leq \varphi(x)$ for all $x \in X$, then there exists a USC function $\chi \colon X \to \mathbb{R}^+$ such that $\chi \leq \varphi - \psi$, the natural bijection h from the graph of φ to the graph of χ is continuous, the restriction $h{\restriction}G_\psi^\varphi \colon G_\psi^\varphi \to G_0^\chi$ is a homeomorphism, and for every $Y \subset X$ such that $\varphi{\restriction}Y$ is a Lelek function with bias $\psi{\restriction}Y$ we have that $\chi{\restriction}Y$ is a Lelek function.*

PROOF. Consider the homeomeomorphism $\alpha \colon [0, \infty) \to [0, 1)$ that is defined by the rule $\alpha(t) = t/(t + 1)$. Note that the derivative $\alpha'(t) \leq 1$ so for each $x \in X$ we have $\alpha(\varphi(x)) - \alpha(\psi(x)) \leq \varphi(x) - \psi(x)$. Thus for the purpose of this proof we may assume that φ and ψ are functions into the interval $[0, 1)$. Let $X \neq \emptyset$, let d be a metric on X and let $U_\varepsilon(x)$ denote the open ε-neighbourhood of $x \in X$. Let ρ be the corresponding max metric on $X \times [0, 1)$: $\rho((x, t), (y, s)) = \max\{d(x, y), |t - s|\}$. Consider the set $A = \{(x, t) : 0 \leq t \leq \psi(x)\}$. Since ψ is USC we have that A is a closed subset of $X \times [0, 1)$. Define for $x \in X$, $\chi(x) = \varphi(x)\rho((x, \varphi(x)), A)$. Note that $\rho((x, \varphi(x)), A) \leq \varphi(x) - \psi(x)$ thus $\chi(x) \leq \varphi(x)(\varphi(x) - \psi(x)) \leq \varphi(x) - \psi(x)$. Clearly, the rule $h(x, \varphi(x)) = (x, \chi(x))$ defines a continuous bijection from the graph of φ to the graph of χ.

First we verify that $\xi(x) = \rho((x, \varphi(x)), A)$ is a USC function which implies that χ is USC as the product of two nonnegative USC functions. Let t be such that

$\xi(x) < t$ and put $\varepsilon = \frac{1}{2}(t - \xi(x))$. Select an $(a, s) \in A$ such that
$$\rho((x, \varphi(x)), (a, s)) = \max\{d(x, a), |\varphi(x) - s|\} < \xi(x) + \varepsilon. \tag{4.8}$$
Since φ is USC there is a $\delta \in (0, \varepsilon)$ such that $\varphi(y) < \varphi(x) + \varepsilon$ whenever $d(x, y) < \delta$. Let y be arbitrary such that $d(x, y) < \delta$. Then by (4.8),
$$d(a, y) \leq d(a, x) + d(x, y) < \xi(x) + \varepsilon + \delta < t. \tag{4.9}$$
If $s \geq \varphi(y)$ then $\rho((y, \varphi(y)), (a, \varphi(y))) = d(y, a) < t$ and since $(a, \varphi(y)) \in A$ we have $\xi(y) < t$. If $s < \varphi(y)$ then by (4.8), $0 < \varphi(y) - s = \varphi(y) - \varphi(x) + \varphi(x) - s < \varepsilon + \xi(x) + \varepsilon = t$. So by (4.9), $\rho((y, \varphi(y)), (a, s)) < t$ and consequently $\xi(y) < t$.

Obviously, $\varphi(x) = \psi(x)$ implies $(x, \varphi(x)) \in A$ and hence $\chi(x) = 0$. If $\chi(x) = 0$ then $\varphi(x) = 0 = \psi(x)$ or $\xi(x) = 0$. If $\xi(x) = 0$ then $(x, \varphi(x)) \in A$ because A is closed. Thus $\varphi(x) \leq \psi(x)$ which means that $\varphi(x) = \psi(x)$. We have shown that $X' = \{x \in X : \chi(x) > 0\} = \{x \in X : \varphi(x) > \psi(x)\}$ and hence $h \restriction G_\psi^\varphi$ is a (continuous) bijection from G_ψ^φ to G_0^χ.

In order to show that $h^{-1} \restriction G_0^\chi$ is continuous consider an $x \in X'$ and a basic neighbourhood B of $(x, \varphi(x))$ in G_ψ^φ. Since φ is USC we may assume that $B = (U_\varepsilon(x) \times (t, 1)) \cap G_\psi^\varphi$ for some $\varepsilon > 0$ and $t \in (0, \varphi(x))$. Put $s = \sqrt{t/\varphi(x)}$ and note that $s < 1$. Since ξ is USC we can find a $\delta \in (0, \varepsilon)$ such that for each $y \in U_\delta(x)$ we have $\xi(y) < \xi(x)/s$. Let $(y, \chi(y))$ be an arbitrary element of $V = (U_\delta(x) \times (s\chi(x), 1)) \cap G_0^\chi$ and note that V is an open neighbourhood of $(x, \chi(x))$. We have
$$\varphi(y) = \frac{\chi(y)}{\xi(y)} > \frac{s\varphi(x)\xi(x)}{\xi(y)} > s^2 \varphi(x) = t \tag{4.10}$$
and hence $h^{-1}(V) \subset B$.

Let $Y \subset X$ be such that $\varphi \restriction Y$ is a Lelek function with bias $\psi \restriction Y$. First, note that $Y' = \{x \in Y : \chi(x) > 0\} = Y \cap X' = \{x \in Y : \psi(x) < \varphi(x)\}$ is dense in Y. Now consider an $\varepsilon > 0$, an $x \in Y'$, and a t such that $0 \leq t \leq \chi(x)$. Define the continuous map $\alpha \colon X \times [0, 1) \to [0, 1)$ by $\alpha(x, s) = s\rho((x, s), A)$. Since $\alpha(x, \psi(x)) = 0$ and $\alpha(x, \varphi(x)) = \chi(x)$ we can find an $s \in [\psi(x), \varphi(x)]$ with $\alpha(x, s) = t$. Since $G_{\psi \restriction Y'}^{\varphi \restriction Y'}$ is dense in $L_{\psi \restriction Y'}^{\varphi \restriction Y'}$ there is a $y \in Y'$ such that $d(y, x) < \varepsilon$ and $|\chi(y) - t| = |\alpha(y, \varphi(y)) - \alpha(x, s)| < \varepsilon$. So (x, t) and $(y, \chi(y))$ are ε-close. $\square$

LEMMA 4.10. *Let X and Z be spaces such that Z is a witness to the almost zero-dimensionality of X. Then the union of the topologies on X and Z is a basis for a topology $\mathcal{T}$ on Z such that the given topology on Z witnesses the almost zero-dimensionality of $(Z, \mathcal{T})$.*

Note that X is an open subspace of $(Z, \mathcal{T})$ and that $\mathcal{T}$ is zero-dimensional at every point of $Z \setminus X$.

PROOF. Note that X is a subset but not necessarily a subspace of Z. Let $\mathcal{B}_1$ and $\mathcal{B}_2$ be countable bases for the topologies on X and Z, respectively. Since $B \cap X$ is open in X whenever $B \in \mathcal{B}_2$ we have that $\mathcal{B}_1 \cup \mathcal{B}_2$ is a basis for the topology $\mathcal{T}$ on Z and that X is an open subspace of $(Z, \mathcal{T})$. Since basic neighbourhoods of points $x \in Z \setminus X$ are elements of $\mathcal{B}_2$ and $\dim Z \leq 0$ we have that $\mathcal{T}$ is zero-dimensional at x. Note that $\mathcal{T}$ is Hausdorff because it contains $\mathcal{B}_2$.

We now verify that Z witnesses the almost zero-dimensionality of $(Z, \mathcal{T})$. If $x \in Z \setminus X$ then basic neighbourhoods of x with respect to $\mathcal{T}$ can be chosen to be clopen subsets of Z. If $x \in X$ then by assumption there is a neighbourhood basis for x in the open set X such that every element is closed in Z. Note that this result also implies that $(X, \mathcal{T})$ is a regular space and hence it is separable metric because there is a countable basis. □

LEMMA 4.11. *Let X be a space and let Z be a zero-dimensional space that contains X as a subset (but not necessarily as a subspace). Then the following statements are equivalent:*

(1) *Z is a witness to the almost zero-dimensionality of X and*
(2) *there exists a USC function $\varphi \colon Z \to \mathbb{I}$ such that $\varphi^{-1}(0) = Z \setminus X$ and the map $h \colon X \to G_0^\varphi$ that is defined by the rule $h(x) = (x, \varphi(x))$ is a homeomorphism.*

In this lemma it does not matter whether the codomain of φ is $\mathbb{I}$ or $\mathbb{R}^+$, but in the applications it is useful to have a bounded φ.

PROOF. To prove (2) ⇒ (1) assume (2). Consider the projection $\pi_1 \colon Z \times \mathbb{I} \to Z$. If O is open in Z then $O \cap X = (\pi_1 \circ h)^{-1}(O)$ is open in X. Let $x \in X$ and let U be a neighbourhood of x in X and note that $\varphi(x) > 0$ because $h(X) = G_0^\varphi$. Then $h(U)$ is a neighbourhood of $h(x)$ in G_0^φ thus there is a neighbourhood V of x in Z and an $\varepsilon \in (0, \varphi(x))$ such that $G_0^\varphi \cap (V \times [\varphi(x) - \varepsilon, \varphi(x) + \varepsilon]) \subset h(U)$. Since φ is USC we can find a (closed) neighbourhood W of x in Z such that $W \subset V$ and $\varphi(y) < \varphi(x) + \varepsilon$ for all $y \in W$. Then $\tilde{W} = \{y \in W : \varphi(y) \geq \varphi(x) - \varepsilon\}$ is closed in Z because φ is USC. If $y \in \tilde{W}$ then $\varphi(y) \geq \varphi(x) - \varepsilon > 0$ so $h(y) \in G_0^\varphi$. Also $y \in W$ and hence $\varphi(y) < \varphi(x) + \varepsilon$ which implies that $h(y) \in h(U)$ and $y \in U$. Thus we have that $\tilde{W}$ is contained in U. Note that $h(\tilde{W}) = G_0^\varphi \cap (W \times [\varphi(x) - \varepsilon, 1])$ is a neighbourhood of $h(x)$ in G_0^φ thus $\tilde{W}$ is a neighbourhood of x in X. We have verified that Z witnesses the almost zero-dimensionality of X.

To prove (1) ⇒ (2) assume (1). With Lemma 4.10 let $\mathcal{T}$ be the topology on Z that is generated by the topologies on X and Z. According to Lemma 3.5 we may assume that the space $\tilde{Z} = (Z, \mathcal{T})$ is the set of end-points of some $\mathbb{R}$-tree $\mathbb{T}$ and that $Z = \mathfrak{e}\mathbb{T}_w$. Let ρ be a convex metric on $\mathbb{T}$ and let $p \in \mathbb{T}$. Define $\varphi \colon Z \to (0, 1]$ by $\varphi(x) = 1/(1 + \rho(p, x))$ for $x \in \mathfrak{e}\mathbb{T}_w$ and note that according to Lemma 3.2 the function φ is USC and $\tilde{Z}$ is homeomorphic to the graph of φ. Thus we have a homeomorphism $f \colon \tilde{Z} \to G_0^\varphi$ that is given by the rule $f(x) = (x, \varphi(x))$.

We define the function $\psi \colon Z \to [0, 1]$ by

(4.11) $$\psi(x) = \mathrm{ext}_Z(\varphi \restriction Z \setminus X).$$

According to Lemma 4.8 ψ is a USC extension of $\varphi \restriction Z \setminus X$. If $x \in X$ then since X is an open subspace of $\tilde{Z}$ and φ is USC there is a neighbourhood U of x in Z and a $t \in (0, \varphi(x))$ such that $(U \times (t, \infty)) \cap G_0^\varphi \subset f(X)$. Consequently, $\varphi(y) \leq t$ for each $y \in U \cap X$ and we have that $\psi(x) \leq t < \varphi(x)$. If $x \in Z \setminus X$ then $\varphi(x) \leq M(\varphi \restriction (U_\varepsilon(x) \setminus X))$ for each $\varepsilon > 0$ and hence $\varphi(x) \leq \psi(x) \leq \varphi(x)$. We have shown that $f(X) = G_\psi^\varphi$. With Lemma 4.9 we can find a USC function $\chi \colon Z \to \mathbb{I}$ such that $\{x \in Z : \chi(x) > 0\} = \{x \in Z : \psi(x) < \varphi(x)\} = X$ and $g(x, \varphi(x)) = (x, \chi(x))$ defines a homeomorphism g from G_ψ^φ to G_0^χ. We define $h = g \circ f$ and note that it is the required homeomorphism from X to G_0^χ. □

REMARK 4.12. Let $\mathcal{T}$ be a witness to the almost zero-dimensionality of some space X and put $Z = (X, \mathcal{T})$. We discuss the relation between the Borel complexities of X and Z. According to Lemma 4.11 we may assume that X is the graph of some USC function $\varphi : Z \to \mathbb{I}$. Let Π_α and Σ_α stand for the productive respectively additive absolute Borel class of rank α. For instance, Π_2 and Σ_2 correspond to the complete respectively σ-compact spaces.

Note that $L(\varphi) = \{(x, r) \in Z \times \mathbb{R} : r \leq \varphi(x)\}$ is closed in $Z \times \mathbb{R}$ because φ is USC. Then $X = L(\varphi) \setminus \bigcup_{n=0}^\infty L(\varphi - 2^{-n})$ is a G_δ-subset of $Z \times \mathbb{I}$. Thus if Z belongs to Π_α for $\alpha \geq 2$ or Σ_α for $\alpha \geq 3$, then X belongs to the same Borel class.

Now let C be a zero-dimensional compactification of Z and let $\psi = \text{ext}_C \varphi$ extend φ with Lemma 4.8.a. According to Remark 2.5 applied to C as a witness to the graph of ψ we have that $Z \in \Pi_{\alpha+1}$ whenever $X \in \Pi_\alpha$ and $Z \in \Sigma_{\alpha+1}$ whenever $X \in \Sigma_\alpha$. For infinite α we even have that Z belongs to the same Borel class as X.

DEFINITION 4.13. If $\varphi : X \to \mathbb{R}^+$ and $\psi : Y \to \mathbb{R}^+$, then $\varphi \times \psi : X \times Y \to \mathbb{R}^+$ is defined by $(\varphi \times \psi)(x, y) = \varphi(x)\psi(y)$.

LEMMA 4.14. *If $\varphi : X \to \mathbb{R}^+$ and $\psi : Y \to \mathbb{R}^+$ are USC functions, then $\varphi \times \psi$ is USC as well and the natural map $h : G_0^\varphi \times G_0^\psi \to G_0^{\varphi \times \psi}$ is a homeomorphism. If, moreover, φ is a Lelek function and $Y' = \{y \in Y : \psi(y) > 0\}$ is dense in Y, then $\varphi \times \psi$ is a Lelek function as well.*

PROOF. Since φ and ψ are nonnegative it is obvious that $\varphi \times \psi$ is USC. The map h is given by $h(x, t, y, s) = (x, y, ts)$ for $(x, t) \in G_0^\varphi$ and $(y, s) \in G_0^\psi$. It is obvious that h is a continuous bijection. It remains to show that h^{-1} is continuous. Let $x \in X$ and $y \in Y$ be such that $\varphi(x) > 0$ and $\psi(y) > 0$. Let $x_1, x_2, \ldots \in X$ and $y_1, y_2, \ldots \in Y$ be such that $\lim_{i \to \infty} x_i = x$, $\lim_{i \to \infty} y_i = y$, and $\lim_{i \to \infty} \varphi(x_i)\psi(y_i) = \varphi(x)\psi(y)$. By the USC property we have $\limsup_{i \to \infty} \varphi(x_i) \leq \varphi(x)$ and $\limsup_{i \to \infty} \psi(y_i) \leq \psi(y)$. Since $\psi(y) > 0$ we can write $\limsup_{i \to \infty} \psi(y_i)/\psi(y) \leq 1$. Observe that

$$(4.12) \quad \begin{aligned} \limsup_{i \to \infty} \varphi(x_i) &\leq \varphi(x) = \lim_{i \to \infty} \frac{\varphi(x_i)\psi(y_i)}{\psi(y)} \\ &\leq \liminf_{i \to \infty} \varphi(x_i) \cdot \limsup_{i \to \infty} \frac{\psi(y_i)}{\psi(y)} \\ &\leq \liminf_{i \to \infty} \varphi(x_i). \end{aligned}$$

So we have $\lim_{i \to \infty} \varphi(x_i) = \varphi(x)$ and by symmetry $\lim_{i \to \infty} \psi(y_i) = \psi(y)$.

Now, let φ be a Lelek function and let Y' be dense in Y. Then $X' = \{x \in X : \varphi(x) > 0\}$ is also dense. Consequently, $X' \times Y' = \{(x, y) : (\varphi \times \psi)(x, y) > 0\}$ is dense in $X \times Y$. Let $(x, y) \in X' \times Y'$, let $t \in [0, \varphi(x)\psi(y)]$, let $U \times V$ be a neighbourhood of (x, y) in $X' \times Y'$, and let $\varepsilon > 0$. Since $0 \leq t/\psi(y) \leq \varphi(x)$ we can find an $x' \in U$ such that $|\varphi(x') - (t/\psi(y))| < \varepsilon/\psi(y)$. We then have that $(x', y) \in U \times V$ and $|(\varphi \times \psi)(x', y) - t| < \varepsilon$. In conclusion, we have that $\varphi \times \psi$ is a Lelek function. □

The following result is already implicitly contained in the papers [**38**, **31**].

THEOREM 4.15. *The following statements about a space X are equivalent:*
(1) *X is almost zero-dimensional,*
(2) *X is homeomorphic to the graph of some USC or LSC function with a domain of dimension at most 0,*

(3) X is imbeddable in complete Erdős space $\mathfrak{E}_c$, and
(4) X is imbeddable in Erdős space $\mathfrak{E}$.

PROOF. (3) $\Rightarrow$ (4) follows from the fact $\mathfrak{E}_c \subset \mathfrak{E}$.
(4) $\Rightarrow$ (1) follows from the fact that almost zero-dimensionality is hereditary.
(1) $\Rightarrow$ (2) follows from Lemma 4.11.
Assume (2) and let $\varphi\colon Z \to (0,1)$ be USC such that $\dim Z = 0$. Let K be a zero-dimensional compactification of Z and consider $\operatorname{ext}_K \varphi \colon K \to \mathbb{I}$. Note that $\{x \in K \colon (\operatorname{ext}_K \varphi)(x) > 0\}$ contains Z and is therefore dense in K. Let C be a Cantor set and let $\psi \colon C \to \mathbb{I}$ be a Lelek function. We may assume that there is a $p \in C$ such that $\psi(p) = 1$. Define $\chi \colon K \times C \to \mathbb{I}$ by $\chi = (\operatorname{ext}_K \varphi) \times \psi$. Note that $\chi(x,p) = (\operatorname{ext}_K \varphi)(x) = \varphi(x)$ for $x \in X$ so G_0^φ and $G_0^{\chi \upharpoonright Z \times \{p\}}$ are identical topological spaces. We have that $K \times C$ is a Cantor set and according to Lemma 4.14, χ is a Lelek function. So according to Kawamura, Oversteegen, and Tymchatyn [**31**] G_0^χ is homeomorphic to $\mathfrak{E}_c$. We have shown that (2) $\Rightarrow$ (3). □

In contrast, the class of totally disconnected spaces has no universal element, see Pol [**39**].

In [**1, 2**] Abry and Dijkstra introduce the notion of an almost n-dimensional space as an extension of almost zero-dimensionality. They prove an n-dimensional version of Theorem 4.15 that includes the construction of higher dimensional analogues of complete Erdős space that are universal spaces for almost n-dimensionality.

REMARK 4.16. The following spaces have a natural representation as in point (2) of the theorem. Let X be a subset of ℓ^p such that X is zero-dimensional in the topology of coordinate-wise convergence and call X equipped with this topology Z. The norm function $N(x) = \|x\|$ is LSC on Z and since the norm function together with the coordinate projections generate the norm topology on ℓ^p we have that X is homeomorphic to the graph of $N \upharpoonright Z$.

COROLLARY 4.17. *Every almost zero-dimensional space has an almost zero-dimensional completion.*

Note that if an almost zero-dimensional space X is σ-compact, then $\dim X \leq 0$.

REMARK 4.18. Let X be an almost zero-dimensional space. Then we can identify X with the graph of an LSC function $\varphi\colon Z \to \mathbb{R}$ with zero-dimensional domain Z. Let $\{A_i \colon i \in \mathbb{N}\}$ be a countable collection of C-sets in X and let $\pi \colon Z \times \mathbb{R} \to Z$ be the projection. We show that we may assume without loss of generality that every $\pi(A_i)$ is closed in Z. Write every $A_i = \bigcap_{j=1}^\infty C_{ij}$, where every C_{ij} is clopen in X. Let $\mathcal{B}$ be the collection of clopen subsets of Z. Strengthen the topology on Z by using

(4.13) $$\mathcal{B}' = \mathcal{B} \cup \{\pi(C_{ij}), Z \setminus \pi(C_{ij}) \colon i, j \in \mathbb{N}\}$$

as a subbasis. Note that now every $\pi(A_i)$ is closed in Z, that Z is still zero-dimensional and separable metric, that φ is still LSC, and that the topology on X is unchanged.

THEOREM 4.19. *A nonempty subset of an almost zero-dimensional space X is a retract of X if and only if it is a C-set in X.*

PROOF. Let $r\colon X \to A$ be a retraction and let x be an arbitrary point in $X \setminus A$. Thus $r(x) \neq x$ and hence by almost zero-dimensionality there is a clopen C in X

with $x \in C$ and $r(x) \notin C$. Consider the clopen neighbourhood $D = C \setminus r^{-1}(C)$ of x and note that $D \cap A = \emptyset$.

For the converse, let A be a nonempty C-set of X. We may by Theorem 4.15 and Remark 4.18 assume that X is the graph of an LSC function $\varphi \colon Z \to \mathbb{I}$ such that Z is zero-dimensional and the image of A under the projection $\pi \colon Z \times \mathbb{I} \to Z$ is closed in Z. Put $A' = \pi(A)$. Let d be a metric on Z and let ρ be a corresponding metric on $Z \times \mathbb{I}$: $\rho((x,s),(y,t)) = d(x,y) + |s-t|$. Since $\dim Z = 0$ we can construct a pairwise disjoint collection $\mathcal{U}$ of nonempty clopen subsets of Z such that $\bigcup \mathcal{U} = Z \setminus A'$ and $\operatorname{diam} U < d(U, A')$ for every $U \in \mathcal{U}$. Select for every $U \in \mathcal{U}$ a $y_U \in U$ such that

$$(4.14) \qquad \varphi(y_U) < \inf\{\varphi(x) \colon x \in U\} + d(U, A')$$

and put $p_U = (y_U, \varphi(y_U)) \in X$. Choose for every $U \in \mathcal{U}$ a $\xi_U \in A$ such that

$$(4.15) \qquad \rho(p_U, \xi_U) \leq 2\rho(p_U, A).$$

We define the retraction $r \colon X \to A$ by

$$(4.16) \qquad r(x, \varphi(x)) = \begin{cases} (x, \varphi(x)), & \text{if } x \in A'; \\ \xi_U, & \text{if } x \in U \in \mathcal{U}. \end{cases}$$

All we need to show is that r is continuous. First note that $r{\restriction}A$ and $r{\restriction}Z \setminus A$ are obviously continuous. So consider a sequence $x_1, x_2, \ldots$ in $Z \setminus A'$ such that $\lim_{i \to \infty} x_i = x \in A'$ and $\lim_{i \to \infty} \varphi(x_i) = \varphi(x)$. For every $i \in \mathbb{N}$ there is a (unique) $U(i) \in \mathcal{U}$ with $x_i \in U(i)$. We have for each $i \in \mathbb{N}$,

$$(4.17) \qquad \begin{aligned} d(y_{U(i)}, x) &\leq d(y_{U(i)}, x_i) + d(x_i, x) \\ &\leq \operatorname{diam} U(i) + d(x_i, x) \\ &\leq d(U(i), A') + d(x_i, x) \\ &\leq 2 d(x_i, x) \end{aligned}$$

thus $\lim_{i \to \infty} y_{U(i)} = x$. Since φ is LSC we have

$$(4.18) \qquad \liminf_{i \to \infty} \varphi(y_{U(i)}) \geq \varphi(x).$$

On the other hand,

$$(4.19) \qquad \begin{aligned} \limsup_{i \to \infty} \varphi(y_{U(i)}) &\leq \limsup_{i \to \infty} (\varphi(x_i) + d(U(i), A')) \\ &\leq \varphi(x) + \lim_{i \to \infty} d(x_i, x) \\ &= \varphi(x). \end{aligned}$$

Thus we have $\lim_{i \to \infty} p_{U(i)} = (x, \varphi(x))$. Note that

$$(4.20) \qquad \begin{aligned} \rho(\xi_{U(i)}, (x, \varphi(x))) &\leq \rho(\xi_{U(i)}, p_{U(i)}) + \rho(p_{U(i)}, (x, \varphi(x))) \\ &\leq 2\rho(p_{U(i)}, A) + \rho(p_{U(i)}, (x, \varphi(x))) \\ &\leq 3\rho(p_{U(i)}, (x, \varphi(x))) \end{aligned}$$

and hence $\lim_{i \to \infty} r(x_i, \varphi(x_i)) = \lim_{i \to \infty} \xi_{U(i)} = (x, \varphi(x))$, as required. □

This theorem generalizes the known result that the retracts of a zero-dimensional space are precisely the nonempty closed subsets. Note that a space is zero-dimensional if and only if every closed set is a C-set. Thus every one-dimensional, almost zero-dimensional space has closed subsets that are not C-sets and hence not retracts of the space. For Erdős space an example would be for instance the unit sphere S. In that space every clopen set that contains the zero vector intersects S so S is no C-set.

The following consequences of Theorem 4.19 are immediate.

COROLLARY 4.20. *Let A be a C-set in an almost zero-dimensional space X. Every clopen subset of A can be extended to a clopen subset of X. If B is a C-set in A then B is also a C-set in X.*

It is shown in Abry, Dijkstra, and van Mill [3] that Theorem 4.19 and Corollary 4.20 are not valid in the class of totally disconnected spaces.

CHAPTER 5

Cohesion

As was mentioned in Chapter 1, Erdős [29] proved that every nonempty clopen subset of $\mathfrak{E}$ is has diameter at least 1. This means that every vector in $\mathfrak{E}$ has a neighbourhood that does not contain any nonempty clopen subsets of $\mathfrak{E}$. This property of $\mathfrak{E}$ turns out to be crucial, and we formalize it as follows.

DEFINITION 5.1. Let X be a space and let $\mathcal{A}$ be a collection of subsets of X. The space X is called $\mathcal{A}$-*cohesive* if every point of the space has a neighbourhood that does not contain nonempty clopen subsets of any element of $\mathcal{A}$. If a space X is $\{X\}$-cohesive then we simply call X *cohesive*.

Thus the prototypical examples of cohesive almost zero-dimensional spaces are $\mathfrak{E}$ and $\mathfrak{E}_c$. In fact, it follows easily from Erdős' proof [29] that the empty set is the only bounded clopen set in these spaces; see also Corollary 8.11. Proposition 4.4 shows that if φ is a Lelek function with complete domain, then G_0^φ is cohesive.

REMARK 5.2. Let X be $\mathcal{A}$-cohesive.

If O is an open subset of X then O is $\{O \cap A : A \in \mathcal{A}\}$-cohesive as follows. Let $x \in O$ and let U be a neighbourhood of x in X that does not contain a nonempty clopen subsets of any $A \in \mathcal{A}$. Select a closed neighbourhood V of x in X that is contained in $U \cap O$. If C is clopen in $A \cap O$ with $A \in \mathcal{A}$ and $C \subset V$, then C is open in A and closed in $V \cap A$ and hence in A as well. Also $C \subset V \subset U$ thus C must be empty.

If Y is any space then $X \times Y$ is $\{A \times B : A \in \mathcal{A} \text{ and } B \subset Y\}$-cohesive. Let $(x, y) \in X \times Y$ and select a neighbourhood U of x in X that does not contain nonempty clopen subsets of any element of $\mathcal{A}$. Let $C \subset U \times Y$ be a clopen subset of $A \times B$ with $A \in \mathcal{A}$ and $B \subset Y$ and assume that $(a, b) \in C$. Then $C \cap (A \times \{b\})$ is a nonempty clopen subset of $A \times \{b\}$ that is also contained in $U \times \{b\}$, in contradiction to the properties of U.

In particular, the product of a cohesive space with any space is cohesive and the concept is open hereditary.

A cohesive space is obviously at least one-dimensional at every point but it is easily seen that the converse is not valid. However, the situation is simple for topological groups:

PROPOSITION 5.3. *A topological group is cohesive if and only if it is not zero-dimensional.*

PROOF. Let $(G, \cdot)$ be a topological group that is not cohesive. Thus there is a point $x \in G$ every neighbourhood of which contains nonempty clopen subsets of G. We may assume that x equals the unit element e. Let U be an arbitrary neighbourhood of e and select a neighbourhood V of e such that $V^{-1} \cdot V \subset U$.

Then there is a nonempty clopen set C with $C \subset V$. Select $x \in C$ and note that $x^{-1} \cdot C$ is a clopen neighbourhood of e that is contained in U. Thus the group G is zero-dimensional at e and hence a zero-dimensional space. □

It is shown by Dijkstra [18] that Proposition 5.3 cannot be extended from topological groups to arbitrary homogeneous spaces.

A *one-point connectification* of a space X is a connected extension Y of the space such that the remainder $Y \setminus X$ is a singleton. It was shown in [33] that the endpoint set of the Lelek fan has a one-point connectification. The relevance of this concept to complete Erdős space was recognized in [31], where it was proved that the endpoint set of the Lelek fan is homeomorphic to $\mathfrak{E}_c$.

PROPOSITION 5.4. *If a space admits a one-point connectification, then it is cohesive. If an almost zero-dimensional space is cohesive, then it admits a one-point connectification.*

PROOF. Let $X \subset Y$ such that $Y \setminus X = \{a\}$ and Y is connected. If $x \in X$ let U be a closed neighbourhood of x in Y that does not contain a. If C is a subset of U that is clopen in X, then C is obviously a clopen subset of Y that does not contain a. Thus C is empty.

Let X be almost zero-dimensional and cohesive. Then we can construct a countable collection $\mathcal{B}$ consisting of subsets of X such that

(1) for every $x \in X$ and every neighbourhood U of x there is a $B \in \mathcal{B}$ such that $x \in \operatorname{int} B \subset B \subset U$,
(2) every $B \in \mathcal{B}$ is an intersection of clopen subsets of X, and
(3) every $B \in \mathcal{B}$ fails to contain nonempty clopen subsets of X.

The combination of the properties (2) and (3) is preserved under finite unions. Let B_1 and B_2 satisfy both (2) and (3). Note that $B_1 \cup B_2 = \bigcap \{C_1 \cup C_2 : C_1, C_2$ clopen and $B_1 \subset C_1, B_2 \subset C_2\}$ thus $B_1 \cup B_2$ also satisfies (2). If C is a nonempty clopen set that is contained in $B_1 \cup B_2$, then C is not contained in B_1 and we pick an $x \in C \setminus B_1$. Select a clopen set D such that $x \notin D$ and $B_1 \subset D$. Then $C \setminus D$ is a clopen nonempty set that is contained in B_2, a contradiction.

Consider the countable set $\mathcal{D} = \{(B_1, B_2) \in \mathcal{B}^2 : B_1 \subset \operatorname{int} B_2\}$ and select for every $D = (B_1, B_2) \in \mathcal{D}$ a continuous function $f_D : X \to \mathbb{I}$ such that $f_D(B_1) \subset \{1\}$ and $f_D(X \setminus B_2) \subset \{0\}$. Let h be the Alexandroff-Urysohn imbedding of X into the Hilbert cube $\mathbb{I}^\mathcal{D}$, given by $h(x)_D = f_D(x)$. Let $Y = h(X) \cup \{\mathbf{0}\}$, where $\mathbf{0}$ represents the element of $\mathbb{I}^\mathcal{D}$ with all coordinates equal to 0. Let C be a clopen subset of Y that does not contain $\mathbf{0}$. Then there exist a finite subset $\{D_1, \ldots, D_n\}$ of $\mathcal{D}$ such that the set $h^{-1}(C)$ is contained in $\bigcup_{i=1}^n f_{D_i}^{-1}((0,1])$. Thus the clopen set $h^{-1}(C)$ is contained in the union of n elements of $\mathcal{B}$ and must be empty. Consequently, $C = \emptyset$ and Y is connected. □

It is shown by Abry, Dijkstra, and van Mill [4] that there are totally disconnected cohesive spaces that do not admit a one-point connectification.

REMARK 5.5. If X is almost zero-dimensional and cohesive and $\mathcal{T}$ is a witness topology on X, then as in the proof of Proposition 5.4 X has a countable covering $\mathcal{B}$ consisting of sets that are closed in $(X, \mathcal{T})$ such that no element of $\mathcal{B}$ contains nonempty sets that are clopen in X. Observe that since $\mathcal{T}$ is zero-dimensional every element of $\mathcal{B}$ has empty interior in $(X, \mathcal{T})$. Thus any witness topology on a cohesive space is of the first category.

Erdős space is obviously an $F_{\sigma\delta}$-subset of Hilbert space thus it is an (absolute) $F_{\sigma\delta}$-space. Since $\{z \in \mathfrak{E} : |z_i| \leq 2^{-i} \text{ for } i \in \omega\}$ is a closed copy of $\mathbb{Q}^\omega$ in $\mathfrak{E}$ we have that $\mathfrak{E}$ is an essential $F_{\sigma\delta}$-space. In fact, it follows that the empty set is the only open subset of $\mathfrak{E}$ that is a $G_{\delta\sigma}$-space. Recall that a space is homeomorphic to $\mathbb{Q}^\omega$ if and only if it is a zero-dimensional, first category $F_{\sigma\delta}$-space with the property that no nonempty open subset is a $G_{\delta\sigma}$-space, see Steel [**41**] or van Engelen [**28**, Theorem A.2.5]. The following proposition follows if we combine these observations with Remarks 4.12 and 5.5.

PROPOSITION 5.6. *Let $\mathfrak{T}$ be a witness topology on $\mathfrak{E}$. Then $(\mathfrak{E}, \mathfrak{T})$ is a $G_{\delta\sigma\delta}$-space but no $G_{\delta\sigma}$-space. If $\mathfrak{T}$ is an $F_{\sigma\delta}$-topology then $(\mathfrak{E}, \mathfrak{T})$ is homeomorphic to $\mathbb{Q}^\omega$.*

REMARK 5.7. Let X be almost zero-dimensional and consider the (zero-dimensional) topology $\mathfrak{T}$ that is generated by *all* clopen subsets of X. The following observation shows why this topology is in general not usable as a witness topology. If X is cohesive then the space $(X, \mathfrak{T})$ has uncountable character at every point.

Let $x \in X$ and let $\{U_i : i \in \mathbb{N}\}$ be a collection of clopen neighbourhoods of x. Let $\mathcal{B} = \{B_i : i \in \mathbb{N}\}$ be a collection that satisfies conditions (1)–(3) as in the proof of Proposition 5.4. We may assume that B_1 contains x. For every $n \in \mathbb{N}$ we can find a point $a_n \in U_n \setminus \bigcup_{i=1}^n B_i$. Let C_n be a clopen subset of X such that $a_n \notin C_n$ and $\bigcup_{i=1}^n B_i \subset C_n$. We define $V = \bigcap_{n=1}^\infty C_n$ and note that V is a closed set that contains B_1 and hence x. If $y \in V$ then there is an $n \in \mathbb{N}$ such that B_n is a neighbourhood of y in X. Observe that $B_n \subset C_k$ for each $k \geq n$ and hence $B_n \cap \bigcap_{i=1}^{n-1} C_i$ is a neighbourhood of y that is contained in V. Thus we have that V is a clopen neighbourhood of x. Note that for any n, $a_n \notin V$ and hence V does not contain U_n, proving that $\{U_n : n \in \mathbb{N}\}$ is no neighbourhood basis of x in $(X, \mathfrak{T})$.

LEMMA 5.8. *Let φ be a USC function from a zero-dimensional space X to $\mathbb{R}^+$ and let $\mathcal{A}$ be a collection of subsets of X such that $\emptyset \notin \mathcal{A}$, G_0^φ is $\{G_0^{\varphi \restriction A} : A \in \mathcal{A}\}$-cohesive, and $A' = \{x \in A : \varphi(x) > 0\}$ is dense in A for each $A \in \mathcal{A}$. Then there exists a USC function $\psi \colon X \to \mathbb{R}^+$ such that $\psi \leq \varphi$, $G_0^\varphi = G_\psi^\varphi$, and for each $A \in \mathcal{A}$ we have that $\varphi \restriction A$ is a Lelek function with bias $\psi \restriction A$.*

PROOF. Let d be a metric on X and let $U_\varepsilon(x)$ denote the open ε-neighbourhood of $x \in X$. We define for $x \in X$ and $\varepsilon > 0$ the following subinterval of $\mathbb{R}^+$:

$$(5.1) \quad J_\varepsilon(x) = \{t \in \mathbb{R}^+ : \text{for each } A \in \mathcal{A}, \text{ the set } U_\varepsilon(x) \times (t, \infty)) \cap G_0^\varphi$$
$$\text{contains no nonempty clopen subsets of } G_0^{\varphi \restriction A}\}.$$

Since φ is USC there exists for each $x \in X$ and $t > \varphi(x)$ an $\varepsilon > 0$ such that $(U_\varepsilon(x) \times (t, \infty)) \cap G_0^\varphi = \emptyset$ and hence $t \in J_\varepsilon(x)$. Note also that if $\varepsilon < \delta$ then $J_\delta(x) \subset J_\varepsilon(x)$. Consequently,

$$(5.2) \quad \psi(x) = \liminf_{\varepsilon \searrow 0} J_\varepsilon(x)$$

is a well-defined function from X to $\mathbb{R}^+$ with the property $\psi(x) \leq \varphi(x)$ for all $x \in X$.

We verify that $G_\psi^\varphi = G_0^\varphi$ which is equivalent to the statement $\{x \in X : \varphi(x) > 0\} = \{x \in X : \psi(x) < \varphi(x)\}$. Let $x \in X$ so be such that $\varphi(x) > 0$. By cohesion and upper semi-continuity there is an $\varepsilon > 0$ such that $(U_\varepsilon(x) \times (\varphi(x) - \varepsilon, \infty)) \cap G_0^\varphi$ contains no nonempty clopen subset of $G_0^{\varphi \restriction A}$ for any $A \in \mathcal{A}$. This means that

$\varphi(x) - \varepsilon \in J_\varepsilon(x)$ and hence $\psi(x) \leq \inf J_\varepsilon(x) \leq \varphi(x) - \varepsilon < \varphi(x)$. Since the other inclusion follows immediately from the fact that $0 \leq \psi$ we are done.

To prove that ψ is USC let $\psi(x) < t$. Then there is an $\varepsilon > 0$ such that $(U_\varepsilon(x) \times (t, \infty)) \cap G_0^\varphi$ contains no nonempty clopen subsets of $G_0^{\varphi \restriction A}$ for any $A \in \mathcal{A}$. So for each $y \in U_\varepsilon(x)$ we have $\psi(y) \leq t$.

Let $A \in \mathcal{A}$. It remains to verify that $\varphi \restriction A$ is a Lelek function with bias $\psi \restriction A$. Since A' is dense in A and equal to $\{x \in A : \psi(x) < \varphi(x)\}$ it suffices to show that $G_{\psi \restriction A}^{\varphi \restriction A}$ is dense in $L_{\psi \restriction A'}^{\varphi \restriction A'}$. Let $x \in A'$, let $t \in (\psi(x), \varphi(x))$, and let $\varepsilon > 0$. There is a $\delta < \varepsilon$ such that $Y = (U_\delta(x) \times (t, \infty)) \cap G_0^\varphi$ contains no nonempty clopen subsets of $G_0^{\varphi \restriction A}$. Select a clopen neighbourhood C of x in X that is contained in $U_\delta(x)$ and consider the set $V = (C \times (t, \infty)) \cap G_0^{\varphi \restriction A}$. Since V is a nonempty subset of both Y and $G_0^{\varphi \restriction A}$ we have that it has boundary points in $G_0^{\varphi \restriction A}$. These boundary points must be of the form (y, t) with $\varphi(y) = t > 0$ and $y \in C \cap A$. Consequently, $(y, \varphi(y))$ is an element of $G_0^{\varphi \restriction A}$ that is sufficiently close to (x, t). □

If we combine Lemma 5.8 with Lemma 4.9, then we obtain

LEMMA 5.9. *Let φ be a USC function from a zero-dimensional space X to $\mathbb{R}^+$ and let $\mathcal{A}$ be a collection of subsets of X such that $\emptyset \notin \mathcal{A}$, G_0^φ is $\{G_0^{\varphi \restriction A} : A \in \mathcal{A}\}$-cohesive, and $A' = \{x \in A : \varphi(x) > 0\}$ is dense in A for each $A \in \mathcal{A}$. Then there exists a USC function $\chi : X \to \mathbb{R}^+$ such that $\chi \leq \varphi$, the natural bijection h from the graph of φ to the graph of χ is continuous, the restriction $h \restriction G_0^\varphi : G_0^\varphi \to G_0^\chi$ is a homeomorphism, and for every $A \in \mathcal{A}$ we have that $\chi \restriction A$ is a Lelek function.*

Lemmas 5.7, 5.8, 5.9, and Remark 5.10 of the preprint version are now Lemmas 4.8, 4.9, 4.11, and Remark 4.12.

PROPOSITION 5.10. *If E is a nonempty, cohesive, almost zero-dimensional space, then there is a Lelek function χ such that E is homeomorphic to G_0^χ and hence E admits a dense imbedding in $\mathfrak{E}_c$.*

PROOF. Assume that E is such a space. With Theorem 4.15 we can find a USC function $\varphi : X \to \mathbb{I}$ such that E is homeomorphic to G_0^φ and $\dim X = 0$. Since E is cohesive we can find with Lemma 5.9 a Lelek function $\chi : X \to \mathbb{I}$ such that G_0^χ is homeomorphic to G_0^φ. Let K be a zero-dimensional compactification of X and note that Lemma 4.8 implies that $\mathrm{ext}_K \chi$ is a Lelek function as well such that G_0^χ is a dense subset of $G_0^{\mathrm{ext}_K \chi}$. Since the domain of $\mathrm{ext}_K \chi$ is compact we have according to Kawamura, Oversteegen, and Tymchatyn [31] that $\mathfrak{E}_c$ is homeomorphic to $G_0^{\mathrm{ext}_K \chi}$. □

REMARK 5.11. The cohesion concept also plays an important role in characterizing complete Erdős space. For instance, the proof above can easily be adapted to show that a nonempty space E is homeomorphic to $\mathfrak{E}_c$ if and only if E is cohesive and there is a topology $\mathcal{T}$ on E that witnesses the almost zero-dimensionality of E such that every point in E has a neighbourhood that is compact in $(E, \mathcal{T})$. This and other characterizations of $\mathfrak{E}_c$ can be found in Dijkstra and van Mill [22].

Theorems 5.13 and 5.16 of the preprint version are now Theorems 4.15 and 4.19.

CHAPTER 6

Unknotting Lelek functions

Let $\varphi\colon X \to \mathbb{R}^+$ and $\psi\colon Y \to \mathbb{R}^+$ be functions. We say that φ and ψ are *m-equivalent* functions if there exists a homeomorphism $h\colon X \to Y$ and a continuous map $\alpha\colon X \to (0,\infty)$ such that $\psi \circ h = \alpha \cdot \varphi$. Note that this is an equivalence relation and that if φ and ψ are m-equivalent, then G_0^φ is homeomorphic to G_0^ψ. Note also that a function that is m-equivalent to a Lelek function is also a Lelek function.

In this section we will prove a Uniqueness Theorem and a Homeomorphism Extension Theorem for Lelek functions with compact domain. Our Uniqueness Theorem is essentially a controlled version of the Characterization Theorem of the Lelek fan that is due to Charatonik [14] and Bula and Oversteegen [12].

LEMMA 6.1. *Let $\varphi\colon C \to \mathbb{R}^+$ and $\psi\colon D \to \mathbb{R}^+$ be Lelek functions with C and D compact metric spaces. If $M(\varphi) = M(\psi)$ and $\varepsilon > 0$, then there is a clopen partition $\mathcal{U}$ of C and a homeomorphism $h\colon C \to D$ such that $\operatorname{mesh}\mathcal{U} < \varepsilon$, $\operatorname{mesh} h[\mathcal{U}] < \varepsilon$, and for each $U \in \mathcal{U}$,*

$$(6.1) \qquad \left|\log \frac{M(\psi \circ h \restriction U)}{M(\varphi \restriction U)}\right| < \varepsilon.$$

PROOF. As noted in Remark 4.6 C and D must be Cantor sets. Since φ and ψ are USC with compact domains we have $\varphi(a) = M(\varphi) = M(\psi) = \psi(b)$ for some $a \in C$, $b \in D$. Select clopen partitions $\mathcal{A} = \{A_0, \ldots, A_m\}$ and $\mathcal{B} = \{B_0, \ldots, B_n\}$ of C respectively D such that $a \in A_0$, $b \in B_0$, $\operatorname{mesh}\mathcal{A} < \varepsilon$ and $\operatorname{mesh}\mathcal{B} < \varepsilon$. For $i \in \{1, \ldots, m\}$ the Lelek property allows us to select distinct points $b_i \in B_0 \setminus \{b\}$ (that approximate b) such that

$$(6.2) \qquad \left|\log \frac{\psi(b_i)}{M(\varphi \restriction A_i)}\right| < \varepsilon/2.$$

Choose disjoint clopen sets $V_1, \ldots, V_m$ in D such that $b_i \in V_i \subseteq B_0 \setminus \{b\}$ and

$$(6.3) \qquad \log \frac{M(\psi \restriction V_i)}{\psi(b_i)} \in [0, \varepsilon/2)$$

by upper semi-continuity. Note that

$$(6.4) \qquad \left|\log \frac{M(\psi \restriction V_i)}{M(\varphi \restriction A_i)}\right| \leq \left|\log \frac{M(\psi \restriction V_i)}{\psi(b_i)}\right| + \left|\log \frac{\psi(b_i)}{M(\varphi \restriction A_i)}\right| < \varepsilon.$$

Conversely, we can find disjoint clopen sets $U_1, \ldots, U_n$ contained in $A_0 \setminus \{a\}$ with

$$(6.5) \qquad \left|\log \frac{M(\psi \restriction B_i)}{M(\varphi \restriction U_i)}\right| = \left|\log \frac{M(\varphi \restriction U_i)}{M(\psi \restriction B_i)}\right| < \varepsilon.$$

Define $\mathcal{U} = \{A_1, \ldots, A_m, U_1, \ldots, U_n, A_0 \setminus \bigcup_{i=1}^n U_i\}$. Let $h\colon C \to D$ be a homeomorphism with $h(A_i) = V_i$ and $h(U_j) = B_j$ for $1 \leq i \leq m$, $1 \leq j \leq n$ and

$h(A_0 \setminus \bigcup_{i=1}^n U_i) = B_0 \setminus \bigcup_{i=1}^m V_i$. Note that

$$(6.6) \qquad \log \frac{M(\psi \restriction B_0 \setminus \bigcup_{i=1}^m V_i)}{M(\varphi \restriction A_0 \setminus \bigcup_{i=1}^n U_i)} = \log \frac{\psi(b)}{\varphi(a)} = 0,$$

so with (6.4) and (6.5) the lemma is proved. $\square$

Remark 6.2 of the preprint version is now Remark 5.2.

THEOREM 6.2 (Uniqueness). *If $\varphi \colon C \to \mathbb{R}^+$ and $\psi \colon D \to \mathbb{R}^+$ are Lelek functions with C and D compact and if $t > |\log(M(\varphi)/M(\psi))|$, then there are a homeomorphism $h \colon C \to D$ and a continuous $\alpha \colon C \to (0, \infty)$ such that $\psi \circ h = \alpha \cdot \varphi$ and $M(\log \circ \alpha) < t$.*

PROOF. Let $\varepsilon = t - |\log(M(\psi)/M(\varphi))| > 0$ and select metrics on C and D that are bounded by 1. We construct by recursion sequences of clopen partitions $\mathcal{U}_0, \mathcal{U}_1, \ldots$ of C and homeomorphisms $h_0, h_1, \ldots$ from C to D for every $n \in \omega$,

(1) if $n \geq 1$ then $\mathcal{U}_n$ refines $\mathcal{U}_{n-1}$,
(2) mesh $\mathcal{U}_n \leq 2^{-n}$,
(3) mesh $h_n[\mathcal{U}_n] \leq 2^{-n}$,
(4) if $n \geq 1$ then $h_n(U) = h_{n-1}(U)$ for each $U \in \mathcal{U}_{n-1}$, and
(5) if $n \geq 1$, $U \in \mathcal{U}_{n-1}$, $V \in \mathcal{U}_n$ such that $V \subset U$, then $|\log(\gamma_V/\gamma_U)| < \varepsilon 2^{-n}$, where

$$(6.7) \qquad \gamma_U = \frac{M(\psi \circ h_{n-1} \restriction U)}{M(\varphi \restriction U)}, \qquad \gamma_V = \frac{M(\psi \circ h_n \restriction V)}{M(\varphi \restriction V)}.$$

Let $h_0 \colon C \to D$ be some homeomorphism and put $\mathcal{U}_0 = \{C\}$. Note that the induction hypotheses are trivially satisfied for $n = 0$. Assume now that h_n and $\mathcal{U}_n$ have been constructed for some $n \in \omega$. Let $U \in \mathcal{U}_n$ and note that $M(\gamma_U \varphi \restriction U) = M(\psi \restriction h_n(U))$ thus we may apply Lemma 6.1 to the pair $\gamma_U \varphi \restriction U$ and $\psi \restriction h_n(U)$ to produce a clopen partition $\mathcal{V}_U$ of U and a homeomorphism $f_U \colon U \to h_n(U)$ such that mesh $\mathcal{V}_U \leq 2^{-n-1}$, mesh $f_U[\mathcal{V}_U] \leq 2^{-n-1}$, and

$$(6.8) \qquad \left| \log \frac{M(\psi \circ f_U \restriction V)}{M(\gamma_U \varphi \restriction V)} \right| < \varepsilon 2^{-n-1}$$

for each $V \in \mathcal{V}_U$. Define

$$(6.9) \qquad \mathcal{U}_{n+1} = \bigcup_{U \in \mathcal{U}_n} \mathcal{V}_U \quad \text{and} \quad h_{n+1} = \bigcup_{U \in \mathcal{U}_n} f_U.$$

Let $V \in \mathcal{V}_U$ and note that

$$(6.10) \qquad \frac{\gamma_V}{\gamma_U} = \frac{M(\psi \circ h_{n+1} \restriction V)}{\gamma_U M(\varphi \restriction V)} = \frac{M(\psi \circ f_U \restriction V)}{M(\gamma_U \varphi \restriction V)}$$

thus hypothesis (5) is satisfied if one also uses formula (6.8). The other induction hypotheses are trivially satisfied and the induction is complete.

Obviously, $h = \lim_{n \to \infty} h_n$ is a homeomorphism $C \to D$. Define for $n \geq 0$ the continuous function $\alpha_n \colon C \to (0, \infty)$ by

$$(6.11) \qquad \alpha_n(x) = \gamma_U \quad \text{for } x \in U \in \mathcal{U}_n.$$

Note that $\alpha_0(x) = M(\psi)/M(\varphi)$ and that

$$(6.12) \qquad |\log(\alpha_n(x)/\alpha_{n-1}(x))| < \varepsilon 2^{-n}$$

for each $x \in C$ and $n \geq 1$. Thus $\log \circ \alpha_0, \log \circ \alpha_1, \ldots$ is a uniform Cauchy sequence of continuous functions into $\mathbb{R}$ and $\alpha = \lim_{n \to \infty} \alpha_n \colon C \to (0, \infty)$ is well-defined and continuous. We have

$$(6.13) \qquad |\log \alpha(x)| < |\log \alpha_0(x)| + \sum_{n=1}^{\infty} \varepsilon 2^{-n} = |\log(M(\psi)/M(\varphi))| + \varepsilon = t$$

for all $x \in C$. Now, let $x \in C$ and select for each n a $U_n \in \mathcal{U}_n$ with $x \in U_n$. Since $h_n(U_n) = h_k(U_n)$ for all $k > n$ we have $h(x) \in h_n(U_n)$. By upper semi-continuity and $\operatorname{diam} U_n \leq 2^{-n}$, $\operatorname{diam} h_n(U_n) \leq 2^{-n}$ we have $\lim_{n \to \infty} M(\varphi \restriction U_n) = \varphi(x)$ and $\lim_{n \to \infty} M(\psi \circ h_n \restriction U_n) = \psi(h(x))$. Thus by (6.11) and (6.7) we have for each $x \in C$,

$$(6.14) \qquad \begin{aligned} \alpha(x)\varphi(x) &= \lim_{n \to \infty} \alpha_n(x) M(\varphi \restriction U_n) \\ &= \lim_{n \to \infty} \gamma_{U_n} M(\varphi \restriction U_n) \\ &= \lim_{n \to \infty} M(\psi \circ h_n \restriction U_n) \\ &= \psi(h(x)), \end{aligned}$$

as required. $\square$

Proposition 6.3 of the preprint version is now Proposition 5.3.

LEMMA 6.3. *Let $\varphi, \psi \colon C \to \mathbb{R}^+$ be Lelek functions with C compact. Let A be a nonempty closed subset of C such that $\varphi \restriction A = \psi \restriction A$, $\{x \in A \colon \varphi(x) > 0\}$ is dense in A, and $G_0^{\varphi \restriction A}$ is nowhere dense in both G_0^{φ} and G_0^{ψ}. If t is a real number with $t > |\log(M(\psi)/M(\varphi))|$, then there exist a homeomorphism $h \colon C \to C$ and a continuous map $\alpha \colon C \to (0, \infty)$ such that $h \restriction A = \operatorname{id}_A$, $\alpha \restriction A = 1_A$, $\psi \circ h = \alpha \cdot \varphi$, and $M(\log \circ \alpha) < t$.*

PROOF. Let d be a metric on C with $\operatorname{diam} C < 1$. We construct a sequence $\mathcal{U}_0, \mathcal{U}_1, \ldots$ of clopen partitions of C with induction hypotheses:

(1) if $n \geq 1$ then $\mathcal{U}_n$ refines $\mathcal{U}_{n-1}$,
(2) if $n \geq 1$, $U \in \mathcal{U}_{n-1}$, and $U \cap A = \emptyset$, then $U \in \mathcal{U}_n$,
(3) if $n \geq 1$, $U \in \mathcal{U}_n$, $U \cap A = \emptyset$, and $U \notin \mathcal{U}_{n-1}$, then $|\log(M(\psi \restriction U)/M(\varphi \restriction U))| < t 2^{-n+1}$,
(4) if $U \in \mathcal{U}_n$ and $U \cap A \neq \emptyset$, then $\operatorname{diam} U < 2^{-n}$ and $|\log(M(\psi \restriction U)/M(\varphi \restriction U))| < t 2^{-n}$.

Put $\mathcal{U}_0 = \{C\}$ and note that hypothesis (4) is satisfied and that the other hypotheses are void.

Assume now that $\mathcal{U}_n$ has been found and let U be an element of $\mathcal{U}_n$ with $U \cap A \neq \emptyset$. So we have $r = |\log(\mu_2/\mu_1)| < t 2^{-n}$, where $\mu_1 = M(\varphi \restriction U)$ and $\mu_2 = M(\psi \restriction U)$. Put $\delta = t 2^{-n} - r$. Since $G_0^{\varphi \restriction A}$ is nowhere dense in both G_0^{φ} and G_0^{ψ} we can select two points p_1 and p_2 in $U \setminus A$ such that

$$(6.15) \qquad \varphi(p_1) > \mu_1 e^{-\delta} \text{ and } \psi(p_2) > \mu_2 e^{-\delta}.$$

Let $\{V_1, \ldots, V_k\}$ be a cover of $U \cap A$ consisting of clopen, pairwise disjoint subsets of C such that $V_i \cap A \neq \emptyset$, $V_i \subset U \setminus \{p_1, p_2\}$, and $\operatorname{diam} V_i < 2^{-n-1}$ for each i. Let $i \in \{1, \ldots, k\}$ and note that by the initial assumptions we have that $s_i = M(\varphi \restriction V_i \cap A) = M(\psi \restriction V_i \cap A) > 0$. Since φ and ψ are USC we can choose a clopen neighbourhood W_i of $V_i \cap A$ in V_i such that

$$(6.16) \qquad W_i \subset \{x \in V_i \colon \varphi(x), \psi(x) < s_i e^{t 2^{-n-1}}\}.$$

Note that both $M(\varphi{\restriction}W_i)$ and $M(\psi{\restriction}W_i)$ are in the interval $[s_i, s_i e^{t2^{-n-1}})$ so we have

(6.17) $\qquad -t2^{-n-1} < \log(M(\psi{\restriction}W_i)/M(\varphi{\restriction}W_i)) < t2^{-n-1}.$

Put $W_0 = U \setminus \bigcup_{i=1}^{k} W_i$ and note that by (6.15),

(6.18) $\qquad \log \dfrac{M(\psi{\restriction}W_0)}{M(\varphi{\restriction}W_0)} \leq \log \dfrac{\mu_2}{\varphi(p_1)} = \log \dfrac{\mu_2}{\mu_1} + \log \dfrac{\mu_1}{\varphi(p_1)} < r + \delta = t2^{-n}.$

and, analogously, $\log(M(\varphi{\restriction}W_0)/M(\psi{\restriction}W_0)) < t2^{-n}$. We define $\mathcal{W}_U = \{W_0, W_1, \ldots, W_k\}$ and

(6.19) $\qquad \mathcal{U}_{n+1} = \{U \in \mathcal{U}_n : U \cap A = \emptyset\} \cup \bigcup \{\mathcal{W}_U : U \in \mathcal{U}_n, U \cap A \neq \emptyset\}.$

and note that the induction hypotheses for $n+1$ are satisfied.

The induction having been completed we define the collection of clopen sets

(6.20) $\qquad \mathcal{V} = \{U : U \in \mathcal{U}_n \text{ for some } n \text{ and } U \cap A = \emptyset\}.$

Hypothesis (4) implies that $\bigcup \mathcal{V} = C \setminus A$. It follows from hypotheses (1) and (2) that $\mathcal{V}$ is a partition of $C \setminus A$. Let U be an element of $\mathcal{V}$ and let n be the first integer such that $U \in \mathcal{U}_n$. Note that $n \geq 1$ since $U \cap A = \emptyset$. By hypothesis (3) we have $|\log(M(\psi{\restriction}U)/M(\varphi{\restriction}U))| < t2^{-n+1}$. Applying Theorem 6.2 to $\varphi{\restriction}U$ and $\psi{\restriction}U$ we find a homeomorphism $f_U : U \to U$ and a continuous map $\beta_U : U \to (0, \infty)$ such that $\psi \circ f_U = \beta_U \cdot \varphi{\restriction}U$ and $M(\log \circ \beta_U) < t2^{-n+1} \leq t$. We define $h : C \to C$ and $\alpha : C \to (0, \infty)$ by

(6.21) $\qquad h(x) = \begin{cases} x, & \text{if } x \in A; \\ f_U(x), & \text{if } x \in U \in \mathcal{V}, \end{cases}$

and

(6.22) $\qquad \alpha(x) = \begin{cases} 1, & \text{if } x \in A; \\ \beta_U(x), & \text{if } x \in U \in \mathcal{V}. \end{cases}$

It is obvious that $h{\restriction}A = \mathrm{id}_A$, $\alpha{\restriction}A = 1_A$, $\psi \circ h = \alpha \cdot \varphi$, and $M(\log \circ \alpha) < t$. It is also obvious that h is a bijection and that $h{\restriction}C \setminus A$ and $\alpha{\restriction}C \setminus A$ are continuous. So let $x \in A$ and let $m \in \mathbb{N}$. Then there is a $U \in \mathcal{U}_m$ that contains x and hence $\mathrm{diam}\, U < 2^{-m}$. Let $y \in U$. If $y \in A$ then $h(y) = y \in U$ and $\alpha(y) = 1 = \alpha(x)$. If $y \in U \setminus A$ then there is a $V \in \mathcal{V}$ that contains y. By hypothesis (1) we have $V \subset U$ and hence $h(y) \in U$ and $|\log(\alpha(y))| < t2^{-n+1}$ for some $n > m$. This proves the continuity of h and α. $\square$

THEOREM 6.4 (Homeomorphism Extension). *Let $\varphi : C \to \mathbb{R}^+$ and $\psi : D \to \mathbb{R}^+$ be Lelek functions with C and D compact. Let $A \subset C$ and $B \subset D$ be closed sets such that $G_0^{\varphi{\restriction}A}$ and $G_0^{\psi{\restriction}B}$ are nowhere dense in G_0^{φ} respectively G_0^{ψ}. Let $h : A \to B$ be a homeomorphism and let $\alpha : A \to (0, \infty)$ be a continuous map such that $\psi \circ h = \alpha \cdot (\varphi{\restriction}A)$. If t is a real number with $t > |\log(M(\psi)/M(\varphi))|$ and $t > M(\log \circ \alpha)$, then there exist a homeomorphism $\tilde{h} : C \to D$ and a continuous map $\tilde{\alpha} : C \to (0, \infty)$ such that $\tilde{h}{\restriction}A = h$, $\tilde{\alpha}{\restriction}A = \alpha$, $\psi \circ \tilde{h} = \tilde{\alpha} \cdot \varphi$, and $M(\log \circ \tilde{\alpha}) < t$.*

PROOF. Let d and ρ be metrics on C respectively D. If $A = \emptyset$ then the theorem is simply Theorem 6.2 thus we may assume that $A \neq \emptyset$. Let $\{q_i : i \in \mathbb{N}\}$ be a countable dense subset of A that is enumerated in such a way that $\{j : q_j = q_i\}$ is infinite for every $i \in \mathbb{N}$. We may assume that α has been continuously extended over C in such a way that $M(\log \circ \alpha) < t$ is still valid. Select for every $i \in \mathbb{N}$ points $a_i \in C \setminus A$ and $b_i \in D \setminus B$ such that $d(q_i, a_i) < 2^{-i}$, $\rho(h(q_i), b_i) < 2^{-i}$, and $\alpha(a_i)\varphi(a_i) = \psi(b_i) > 0$, as follows. First use the fact that ψ is Lelek to find a point $b'_i \in D \setminus B$ with $\rho(h(q_i), b'_i) < 2^{-i}$ and $\psi(b'_i) > 0$. Since $\alpha \cdot \varphi$ is also Lelek we can choose an $a_i \in C \setminus A$ such that $d(q_i, a_i) < 2^{-i}$ and $0 < \alpha(a_i)\varphi(a_i) < \psi(b'_i)$. Finally, using Proposition 4.4 and again the Lelek property of ψ, we find a $b_i \in D \setminus B$ close enough to b'_i such that $\rho(h(q_i), b'_i) < 2^{-i}$ and $\alpha(a_i)\varphi(a_i) = \psi(b_i)$. We can easily arrange that all the a_i's and b_i's are distinct. We define $\hat{A} = A \cup \{a_i : i \in \mathbb{N}\}$, $\hat{B} = B \cup \{b_i : i \in \mathbb{N}\}$, $\hat{\alpha} = \alpha\!\upharpoonright\!\hat{A}$, and the homeomorphism $\hat{h} \colon \hat{A} \to \hat{B}$ by $\hat{h}\!\upharpoonright\!A = h$ and $h(a_i) = b_i$ for $i \in \mathbb{N}$. Obviously, $\psi \circ \hat{h} = \hat{\alpha} \cdot \varphi\!\upharpoonright\!\hat{A}$. Note that every point $(a_i, \varphi(a_i))$ is an isolated point of $G_0^{\varphi\upharpoonright\hat{A}}$ but not isolated in G_0^{φ} so $G^{\varphi\upharpoonright\hat{A}}$ is just as $G_0^{\varphi\upharpoonright A}$ nowhere dense in G_0^{φ}. Analogously, we have that also $G_0^{\psi\upharpoonright\hat{B}}$ is nowhere dense in G_0^{ψ}.

The preceding paragraph shows that we may assume without loss of generality that $\{x \in A : \varphi(x) > 0\}$ is dense in A. Put

(6.23) $$\delta = t - \max\{M(\log \circ \alpha), |\log(M(\psi)/M(\varphi))|\}$$

and select a $b \in D \setminus B$ such that $\psi(b) > M(\psi)e^{-\delta}$. Choose a homeomorphism $h_1 \colon C \to D$ and a continuous map $\alpha_1 \colon C \to (0, \infty)$ such that $h_1\!\upharpoonright\!A = h$, $\alpha_1\!\upharpoonright\!A = \alpha$, and $M(\log \circ \alpha_1) = M(\log \circ \alpha)$. Define $\xi = (1/\alpha_1) \cdot (\psi \circ h_1)$ and note that ξ is a Lelek function on C which coincides with φ on A. Using the fact that ξ is USC we find a clopen neighbourhood U of A such that $h_1^{-1}(b) \notin U$ and $M(\xi\!\upharpoonright\!U) < M(\varphi)e^{\delta}$. We define the continuous function $\alpha_2 \colon C \to (0, \infty)$ by

(6.24) $$\alpha_2(x) = \begin{cases} \alpha_1(x), & \text{if } x \in U; \\ M(\psi)/M(\varphi), & \text{if } x \in C \setminus U. \end{cases}$$

We have

(6.25) $$M(\log \circ \alpha_2) \leq \max\{M(\log \circ \alpha_1), |\log(M(\psi)/M(\varphi))|\} = t - \delta.$$

Then $\chi = (1/\alpha_2) \cdot (\psi \circ h_1)$ is a Lelek function on C which coincides with ξ on U. Note that $\chi\!\upharpoonright\!A = \xi\!\upharpoonright\!A = \varphi\!\upharpoonright\!A$ and that $G_0^{\chi\upharpoonright A}$ is nowhere dense in G_0^{χ} just as $G_0^{\psi\upharpoonright B}$ is nowhere dense in G_0^{ψ}. Observe that

(6.26) $$M(\chi\!\upharpoonright\!U) = M(\xi\!\upharpoonright\!U) < M(\varphi)e^{\delta},$$

(6.27) $$M(\chi\!\upharpoonright\!C \setminus U) \leq \frac{M(\varphi)}{M(\psi)} M(\psi \circ h_1) = M(\varphi),$$

(6.28) $$M(\chi) \geq \chi(h_1^{-1}(b)) = \frac{M(\varphi)}{M(\psi)} \psi(b) > M(\varphi)e^{-\delta},$$

and hence $|\log(M(\chi)/M(\varphi))| < \delta$. According to Lemma 6.3 there exist a homeomorphism $h_2 \colon C \to C$ and a continuous map $\beta \colon C \to (0, \infty)$ such that $h_2\!\upharpoonright\!A = \mathrm{id}_A$, $\beta\!\upharpoonright\!A = 1_A$, $\chi \circ h_2 = \beta \cdot \varphi$, and $M(\log \circ \beta) < \delta$. We define the homeomorphism $\tilde{h} \colon C \to D$ and the continuous map $\tilde{\alpha} \colon C \to (0, \infty)$ by $\tilde{h} = h_1 \circ h_2$ and $\tilde{\alpha} = (\alpha_2 \circ h_2) \cdot \beta$. We have

(6.29) $$\tilde{h}\!\upharpoonright\!A = h_1 \circ h_2\!\upharpoonright\!A = h_1\!\upharpoonright\!A = h,$$

(6.30) $$\tilde{\alpha}\!\upharpoonright\!A = (\alpha_2 \circ h_2\!\upharpoonright\!A) \cdot (\beta\!\upharpoonright\!A) = (\alpha_2\!\upharpoonright\!A) \cdot 1_A = \alpha_1\!\upharpoonright\!A = \alpha,$$

(6.31) $$\psi \circ \tilde{h} = \psi \circ h_1 \circ h_2 = (\alpha_2 \cdot \chi) \circ h_2 = (\alpha_2 \circ h_2) \cdot \beta \cdot \varphi = \tilde{\alpha} \cdot \varphi,$$

(6.32) $$M(\log \circ \tilde{\alpha}) \leq M(\log \circ \alpha_2) + M(\log \circ \beta) < (t - \delta) + \delta = t$$

and the proof is complete. $\square$

CHAPTER 7

Extrinsic characterizations of Erdős space

In this chapter we present two characterizations of Erdős space in terms of imbeddings of the space into graphs of Lelek functions.

DEFINITION 7.1. Let X be a space. We call a system $(X_s)_{s\in T}$ a *Sierpiński stratification* of X if T is a nonempty tree over a countable alphabet and X_s is a closed subset of X for each $s \in T$ such that:
  i. $X_\emptyset = X$ and $X_s = \bigcup\{X_t : t \in \mathrm{succ}(s)\}$ for all $s \in T$, and
  ii. if $\sigma \in [T]$ then the sequence $X_{\sigma\restriction 0}, X_{\sigma\restriction 1}, \ldots$ converges to a point $x_\sigma \in X$.

Recall that Sierpiński [40] has shown that a space is an $F_{\sigma\delta}$-space if and only if it admits a Sierpiński stratification and that van Engelen [28, Theorem A.1.6] has shown that a zero-dimensional space X is homeomorphic to $\mathbb{Q}^\omega$ if there exists a Sierpiński stratification $(X_s)_{s\in T}$ of X such that X_t is nowhere dense in X_s whenever $t \in \mathrm{succ}(s)$. Our characterizations of $\mathfrak{E}$ were inspired by these results.

DEFINITION 7.2. SLC is the class of all pairs (φ, X) such that $\varphi\colon C \to \mathbb{R}^+$ is a USC function with a zero-dimensional compact domain that contains X for which there exist a nonempty tree T over a countable set and closed subsets X_s of C for each $s \in T$ such that:
  (1) $X_\emptyset = C$ and $X_t \subset X_s$ whenever $s \prec t$ and $s, t \in T$,
  (2) for each $s \in T$ and we have $X_s \cap X \subset \bigcup\{X_t : t \in \mathrm{succ}(s)\}$,
  (3) if $\sigma \in [T]$ then $\bigcap_{k=0}^\infty X_{\sigma\restriction k}$ is a singleton $\{x_\sigma\} \subset X$,
  (4) for each $s \in T$ and $t \in \mathrm{succ}(s)$ we have that $G_0^{\varphi\restriction X_t}$ is nowhere dense in $G_0^{\varphi\restriction X_s}$, and
  (5) for each $s \in T$, $\bigcup\{G_0^{\varphi\restriction X_t} : t \in \mathrm{succ}(s)\}$ is dense in $L_0^{\varphi\restriction X_s}$.

(SLC stands for Sierpiński-Lelek-compact.)

We illustrate this definition with an example. Let K be a Cantor set in $\mathbb{R}$ that contains 0 and let A be a countable dense subset of K. Put $X = A^\omega$ and $C = K^\omega$. Let $p \geq 1$ and let $\|\cdot\|$ denote the p-norm on $\mathbb{R}^\omega$. We define $\varphi\colon C \to \mathbb{I}$ by $\varphi(z) = 1/(1+\|z\|)$. If we put $T = A^{<\omega}$ and $X_{a_0\ldots a_{k-1}} = \{a_0\} \times \cdots \times \{a_{k-1}\} \times K \times K \times \cdots$, then it is not hard to see that (φ, X) is an element of SLC, cf. Proposition 8.12.

LEMMA 7.3. *If $(\varphi, X) \in$ SLC then there are a tree T and a system $(X_s)_{s\in T}$ as in Definition 7.2 with the following additional properties: every X_s is nonempty, T is the Baire tree $\mathbb{N}^{<\omega}$, and*
  (6) *for all $s, t \in T$ with $|s| = |t|$ we have $s = t$ or $X_s \cap X_t = \emptyset$.*

PROOF. Let φ, X, T, and $(X_s)_{s\in T}$ be given as in Definition 7.2. First note that we can delete any node s from T with the property $X_s = \emptyset$ without affecting the properties (1)–(5). Now if $s \in T$ then $X_s \neq \emptyset$ and hence $L_0^{\varphi\restriction X_s} \neq \emptyset$. By condition

(5) we have that there is an immediate successor t of s such that $G_0^{\varphi \upharpoonright X_t} \ne \emptyset$ and hence $G_0^{\varphi \upharpoonright X_s} \ne \emptyset$. Conditions (4) and (5) combined now show that there are infinitely many $t' \in \mathrm{succ}(s)$. By relabelling we can arrange that $T = \mathbb{N}^{<\omega}$.

We will now verify that we can make the X_s disjoint. Let A be the countable set $\{(i,s) \in \mathbb{N} \times \mathbb{N}^{<\omega} : |s| = i-1\}$. Let $\pi \colon A \to \mathbb{N}$ be the projection $\pi(i,s) = i$ and let $\pi \colon A^{<\omega} \cup A^\omega \to \mathbb{N}^{<\omega} \cup \mathbb{N}^\omega = T \cup [T]$ also denote the induced projection. We will construct by recursion with respect to the length l of strings from $(X_s)_{s \in T}$ a new system $(Y_s)_{s \in A^{<\omega}}$ that satisfies the following hypotheses for $l \in \omega$:

  (a) if $s, t \in A^{<\omega}$ and $|s| = |t| = l$, then $s = t$ or $Y_s \cap Y_t = \emptyset$,
  (b) if $s \in A^{<\omega}$ and $|s| = l$, then Y_s is a clopen subset of $X_{\pi(s)}$.

We begin with $Y_\emptyset = X_\emptyset = C$. For every $s \in T$ we find a pairwise disjoint collection $\{C_s^i : i \in \mathbb{N}\}$ of clopen subsets of C such that its union equals $C \setminus X_s$. Let us assume that Y_s has been found for $s \in A^{<\omega}$ with $|s| = l$ and let $a = (i, n_1 n_2 \ldots n_{i-1}) \in A$. We define

$$(7.1) \qquad Y_{s^\frown a} = X_{\pi(s)^\frown i} \cap Y_s \cap \bigcap_{j=1}^{i-1} C_{\pi(s)^\frown j}^{n_j}.$$

Note that hypothesis (b) is satisfied because Y_s is clopen in $X_{\pi(s)}$ which contains $X_{\pi(s)^\frown i}$. For hypothesis (a) consider $s, t \in A^{<\omega}$ with $|s| = |t| = l$ and $a, b \in A$ with $s^\frown a \ne t^\frown b$. If $s \ne t$ then by hypothesis $Y_s \cap Y_t = \emptyset$ and hence $Y_{s^\frown a} \cap Y_{t^\frown b} = \emptyset$. So we may assume that $s = t$. Let $a = (i, n_1 \ldots n_{i-1})$ and $b = (j, k_1 \ldots k_{j-1})$. If $i < j$ then $Y_{s^\frown b}$ is contained in $C_{\pi(s)^\frown i}^{k_i}$, which set is disjoint from $X_{\pi(s)^\frown i}$ and hence from $Y_{s^\frown a}$. So we may put $i = j$ and hence there is an m such that $n_m \ne k_m$. The desired conclusion now follows from the fact

$$(7.2) \qquad Y_{s^\frown a} \cap Y_{s^\frown b} \subset C_{\pi(s)^\frown m}^{n_m} \cap C_{\pi(s)^\frown m}^{k_m} = \emptyset.$$

We now put $T' = \{s \in A^{<\omega} : Y_s \ne \emptyset\}$. Since by the definition $Y_t \subset Y_s$ whenever $t \in \mathrm{succ}(s)$ we have that T' is a tree and we also have condition (1) for the system $(Y_s)_{s \in T'}$. Condition (6) follows from hypothesis (a) and conditions (4) and (3) follow from hypothesis (b) and, for (3), compactness. We verify that for every $s \in A^{<\omega}$,

$$(7.3) \qquad \bigcup \{Y_t : t \in \mathrm{succ}(s)\} = Y_s \cap \bigcup \{X_{\pi(t)} : t \in \mathrm{succ}(s)\}.$$

Since the other direction is trivial it suffices to prove that every element x of the right hand side of (7.3) is contained in the left hand side. Choose the lowest index i such that $x \in X_{\pi(s)^\frown i}$. So for every $j < i$, $x \notin X_{\pi(s)^\frown j}$ and hence there is an $n_j \in \mathbb{N}$ with $x \in C_{\pi(s)^\frown j}^{n_j}$. Putting $a = (i, n_1 \ldots n_{i-1})$ we find that $x \in Y_{s^\frown a}$.

With hypothesis (b) we may conclude from (7.3) that condition (5) is satisfied. For condition (2) consider an $s \in T'$. Since $X_{\pi(s)}$ satisfies condition (2) we have, again by (7.3),

$$(7.4) \qquad \begin{aligned} Y_s \cap X &= Y_s \cap X \cap X_{\pi(s)} \\ &\subset Y_s \cap \bigcup \{X_{\pi(t)} : t \in \mathrm{succ}(s)\} \\ &= \bigcup \{Y_t : t \in \mathrm{succ}(s)\}. \end{aligned}$$

Thus Y_s also satisfies condition (2). Finally, as argued above we can replace T' by $\mathbb{N}^{<\omega}$. $\square$

REMARK 7.4. Let $(\varphi, X) \in \mathsf{SLC}$ with a system $(X_s)_{s \in T}$. It follows from conditions (1)–(3) that $(X_s \cap X)_{s \in T}$ is a Sierpiński stratification of X and that $X = \{x_\sigma : \sigma \in [T]\}$. Note that it follows from condition (5) that every $\varphi \restriction X_s$ is a Lelek function. It is also easily seen that if h is a homeomorphism from the domain C of φ to another Cantor set D and $\alpha : C \to (0, \infty)$ is continuous, then $((\alpha \cdot \varphi) \circ h^{-1}, h(X))$ is also in SLC. Finally, conditions (1) and (6) imply that if $X_s \cap X_t \neq \emptyset$ then $X_t \subset X_s$ and $s \prec t$ or $X_s \subset X_t$ and $t \prec s$.

THEOREM 7.5. *Let $(\varphi, X), (\psi, Y) \in \mathsf{SLC}$. Then there exists a homeomorphism f from the domain C of φ to the domain D of ψ and a continuous map $\beta : C \to (0, \infty)$ such that $f(X) = Y$ and $\psi \circ f = \beta \cdot \varphi$ (and hence $G_0^{\varphi \restriction X}$ is homeomorphic to $G_0^{\psi \restriction Y}$).*

REMARK 7.6. Let $(\varphi, X), (\psi, Y) \in \mathsf{SLC}$, let $T = \mathbb{N}^{<\omega}$, and let $(X_s)_{s \in T}$ respectively $(Y_s)_{s \in T}$ be systems of nonempty closed sets in C respectively D as in Definition 7.2 that also satisfy condition (6) of Lemma 7.3. Van Engelen's proof [**28**, pp. 115–120] of the characterization of $\mathbb{Q}^\omega$ in terms of Sierpiński stratifications shows that there exists a homeomorphism $h : C \to D$ with $h(X) = Y$. This homeomorphism will in general not correspond to an m-equivalence between the functions φ and ψ. In order to get a continuous $\beta : C \to (0, \infty)$ such that $\tilde{\psi} \circ h = \beta \cdot \tilde{\varphi}$ and $h(X) = Y$ we need to add an additional ingredient to van Engelen's construction in the form of the Homeomorphism Extension Theorem for Lelek functions (Theorem 6.4).

PROOF. Let $(\varphi, X), (\psi, Y), (X_s)_{s \in T}$, and $(Y_s)_{s \in T}$ be as in Remark 7.6. With the Uniqueness Theorem for Lelek functions (Theorem 6.2) we may assume that $C = D$ and $\psi = \varphi$. Choose a metric d on C with $\operatorname{diam} C < 1$. Split ω into two infinite sets E and F. If $n \in \omega$ then $E(n)$ denotes the set $\{i \in E : i < n\}$. Similarly for F. Let $\tau_E : E \to T$ be a bijection that is *monotone*: if $n, k \in E$ then $\tau_E(n) \prec \tau_E(k)$ implies $n \leq k$. Let $\tau_F : F \to T$ also be a monotone bijection and let $\tau : \omega \to T$ be the function $\tau_E \cup \tau_F$. Note that monotonicity implies that $\tau(0) = \emptyset$.

By induction on n we will construct clopen partitions $\mathcal{U}_0, \mathcal{U}_1, \ldots$ of C, homeomorphisms $h_0, h_1, \ldots : C \to C$, and continuous maps $\beta_0, \beta_1, \ldots : C \to (0, \infty)$ such that, using the notation $f_n = h_n \circ \cdots \circ h_0$, we have for each $n \geq 0$ and $U \in \mathcal{U}_n$,

(1) $\operatorname{mesh} \mathcal{U}_n < 2^{-n}$,
(2) $\mathcal{U}_n$ refines $\mathcal{U}_{n-1}$ if $n \geq 1$,
(3) $h_n(U) = U$,
(4) if $m \in E(n)$ then $f_n(X_{\tau(m)}) \cap U = f_m(X_{\tau(m)}) \cap U$,
(5) if $k \in F(n)$ then $f_n^{-1}(Y_{\tau(k)} \cap U) = f_k^{-1}(Y_{\tau(k)} \cap U)$,
(6) if $n \in E$ and $f_n(X_{\tau(n)}) \cap U \neq \emptyset$, then there exists a $k \in F$ such that $|\tau(n)| = |\tau(k)|$ and $f_n(X_{\tau(n)}) \cap U = Y_{\tau(k)} \cap U$,
(7) if $n \in F$ and $Y_{\tau(n)} \cap U \neq \emptyset$, then there exists an $m \in E$ such that $|\tau(n)| = |\tau(m)|$ and $Y_{\tau(n)} \cap U = f_n(X_{\tau(m)}) \cap U$.
(8) $|\log(\beta_n(x)/\beta_{n-1}(x))| < 2^{-n}$ for $n \geq 1$ and each $x \in f_n^{-1}(U)$,
(9) $\varphi \restriction U = (\beta_n \cdot \varphi) \circ f_n^{-1} \restriction U$.

Put $\mathcal{U}_0 = \{C\}$, let h_0 be the identity map id_C, and let β_0 be the constant map 1_C. Note that for the case $n = 0$ the induction hypotheses are trivially satisfied.

Observe that if $f_n(X_{\tau(n)}) \cap U \neq \emptyset$, where $U \in \mathcal{U}_n$, then the natural number k promised in hypothesis (6) is unique by condition (6) in Lemma 7.3. Similarly for

m in (7). Suppose for a moment that we completed the construction. Hypotheses (1), (2), and (3) imply that $f = \lim_{n\to\infty} f_n$ exists and is a homeomorphism of C. We claim that $f(X) = Y$. To this end, let $x \in X$ be arbitrary. There is a unique $\sigma \in [T] = \mathbb{N}^\omega$ such that $x \in \bigcap_{k=0}^\infty X_{\sigma\restriction k}$. For every i let $n_i \in E$ be such that $\tau(n_i) = \sigma\restriction i$ and let U_i be the unique element of $\mathcal{U}_{n_i}$ that contains $f_{n_i}(x)$. By hypothesis $(6)_{n_i}$ there is a $k_i \in F$ such that $k = |\tau(n_i)| = |\tau(k_i)|$ and

$$(7.5) \qquad f_{n_i}(x) \in f_{n_i}(X_{\tau(n_i)}) \cap U_i = Y_{\tau(k_i)} \cap U_i.$$

So by (3) and (4) we have for every j that

$$(7.6) \qquad f_{n_i+j}(x) \in f_{n_i+j}(X_{\tau(n_i)}) \cap U_i = f_{n_i}(X_{\tau(n_i)}) \cap U_i = Y_{\tau(k_i)} \cap U_i,$$

which means that $f(x) \in Y_{\tau(k_i)}$. Since $|\tau(k_i)| = i$ for every i, this proves that $f(x) \in \bigcap_{k=0}^\infty Y_{\tau(m_k)} \subset Y$. Hence $f(X) \subset Y$, and, by symmetry, $f^{-1}(Y) \subset X$.

It is obvious that hypothesis (8) implies that $(\log \circ \beta_n)_n$ is a uniform Cauchy sequence and hence $\beta = \lim_{n\to\infty} \beta_n \colon C \to (0,\infty)$ exists and is continuous. Observe that by uniform convergence of $(f_n)_n$ and $(\beta_n)_n$ we have that $\lim_{n\to\infty} f_n^{-1} = f^{-1}$ and $\lim_{n\to\infty} \beta_n \circ f_n^{-1} = \beta \circ f^{-1}$. Let $x \in C$ and note that we have by upper semi-continuity of φ and hypothesis (9):

$$(7.7) \begin{aligned} \beta(x)\varphi(x) &= \lim_{n\to\infty} \beta_n(x)\varphi(x) = \lim_{n\to\infty} \varphi(f_n(x)) \\ &\leq \varphi(f(x)) = \lim_{n\to\infty} \varphi(f_n(f_n^{-1}(f(x)))) \\ &= \lim_{n\to\infty} \beta_n(f_n^{-1}(f(x)))\varphi(f_n^{-1}(f(x))) \\ &\leq \beta(x)\varphi(x). \end{aligned}$$

So we have $\varphi \circ f = \beta \cdot \varphi$.

It remains to perform the induction. Suppose that for some n the partition $\mathcal{U}_n$, the homeomorphism h_n, and the map β_n have been found. Because of the symmetry between X and Y we may assume without loss of generality that $n+1 \in E$.

Select a clopen partition $\mathcal{V}$ of C that refines $\mathcal{U}_n$ and with the property mesh $\mathcal{V} < 2^{-n-1}$. Consider an arbitrary element V of $\mathcal{V}$ such that $A_V = f_n(X_{\tau(n+1)}) \cap V \neq \emptyset$. Note that $\varphi\restriction A_V$ is a Lelek function and hence $M(\varphi\restriction A_V) > 0$. Since φ is USC we have that $\{x \in V \colon \varphi(x) < M(\varphi\restriction A_V)e^{2^{-n-1}}\}$ is an open neighbourhood of A_V. We now select a clopen subset U_V of V containing A_V such that

$$(7.8) \qquad U_V \subset \{x \in V \colon \varphi(x) < M(\varphi\restriction A_V)e^{2^{-n-1}}\},$$
$$(7.9) \qquad f_n(X_{\tau(m)}) \cap U_V = \emptyset$$

whenever $m \in E(n+1)$ with $f_n(X_{\tau(n+1)} \cap X_{\tau(m)}) \cap V = \emptyset$,

and

$$(7.10) \qquad Y_{\tau(m)} \cap U_V = \emptyset$$

whenever $m \in F(n+1)$ with $f_n(X_{\tau(n+1)}) \cap Y_{\tau(m)} \cap V = \emptyset$.

Define

$$(7.11) \qquad \mathcal{U}_{n+1} = (\{V \colon V \in \mathcal{V}, A_V = \emptyset\} \cup \{U_V, V \setminus U_V \colon V \in \mathcal{V}, A_V \neq \emptyset\}) \setminus \{\emptyset\}$$

and note that this collection satisfies hypotheses $(1)_{n+1}$ and $(2)_{n+1}$.

Let U be an arbitrary element of $\mathcal{U}_{n+1}$ and put

$$(7.12) \qquad A = f_n(X_{\tau(n+1)}) \cap U.$$

We will define h_{n+1} and β_{n+1} by determining $h_{n+1}\!\restriction\! U$ and $\beta_{n+1}\!\restriction\! f_n^{-1}(U)$. We consider two cases.

Case I: either $A = \emptyset$ or there is a $k \in F(n+1)$ with

(7.13) $\qquad f_n(X_{\tau(n+1)}) \cap U = Y_{\tau(k)} \cap U \quad \text{and} \quad |\tau(n+1)| = |\tau(k)|.$

In this case we put $h_{n+1}\!\restriction\! U = \mathrm{id}_U$ and $\beta_{n+1}(x) = \beta_n(x)$ for each $x \in f_n^{-1}(U)$. This definition trivially satisfies the hypotheses (3)–(9) for $n+1$ and U.

Case II: all situations that are not covered by Case I. Put $l = |\tau(n+1)|$. Note that $l > 0$ because if $l = 0$ then $X_{\tau(n+1)} = C$ and we are in Case I. Since τ_E is a monotone bijection there are unique integers $\nu_0 < \nu_1 < \cdots < \nu_l$ in E such that $\tau(\nu_i) = \tau(n+1)\!\restriction\! i$ for $0 \leq i \leq l$. Let $0 \leq i < l$ and let W be the element of $\mathcal{U}_n$ that contains U and note that $f_n(X_{\tau(\nu_i)}) \cap W$ contains the nonempty set $f_n(X_{\tau(n+1)}) \cap U$. Thus by hypothesis (6) there is a $\kappa_i \in F$ with $|\tau(\kappa_i)| = |\tau(\nu_i)| = i$ and $f_n(X_{\tau(\nu_i)}) \cap W = Y_{\tau(\kappa_i)} \cap W$. Thus $f_n(X_{\tau(\nu_i)}) \cap U = Y_{\tau(\kappa_i)} \cap U$ and $\tau(\kappa_i) \prec \tau(\kappa_{i+1})$ for every $i < l$. Select with condition (5) in Definition 7.2 a $j \in \mathbb{N}$ such that there is a $b \in Y_{\tau(\kappa_{l-1})^\frown j} \cap U$ with the property $\varphi(b) > M(\varphi\!\restriction\! A) e^{-2^{-n-1}}$. Let $\kappa_l \in F$ be such that $\tau(\kappa_l) = \tau(\kappa_{l-1})^\frown j$ and put $B = Y_{\tau(\kappa_l)} \cap U$. By the way that $\mathcal{U}_{n+1}$ was obtained from $\mathcal{V}$ (see (7.8)) we have that $|\log(M(\varphi\!\restriction\! A)/M(\varphi\!\restriction\! B))| < 2^{-n-1}$. Note that if the intersection of any X_s or Y_s with U is nonempty, then $\varphi\!\restriction\! X_s \cap U$ respectively $\varphi\!\restriction\! Y_s \cap U$ is a Lelek function. Thus with Theorem 6.2 we can find a homeomorphism $g\colon A \to B$ and a continuous map $\alpha\colon A \to (0, \infty)$ such that $\varphi \circ g = \alpha \cdot (\varphi\!\restriction\! A)$ and $M(\log \circ \alpha) < 2^{-n-1}$. Note that $G_0^{\varphi\restriction A}$ is nowhere dense in $G_0^{\varphi\restriction f_n(X_{\tau(\nu_{l-1})}) \cap U} = G_0^{\varphi\restriction Y_{\tau(\kappa_{l-1})} \cap U}$ by condition (4) in Definition 7.2. Also $G_0^{\varphi\restriction Y_{\tau(\kappa_i)} \cap U}$ is nowhere dense in $G_0^{\varphi\restriction Y_{\tau(\kappa_{i-1})} \cap U}$ for $0 < i < l$ and $|\log(M(\varphi\!\restriction\! U)/M(\varphi\!\restriction\! A))| < 2^{-n-1}$ by the construction of $\mathcal{U}_{n+1}$. Since $Y_{\tau(\kappa_0)} = Y_\emptyset = C$ we can now use the Homeomorphism Extension Theorem for Lelek functions (Theorem 6.4) recursively to find a homeomorphism $\tilde{g}\colon U \to U$ and a continuous $\tilde{\alpha}\colon U \to (0, \infty)$ such that $\tilde{g}\!\restriction\! A = g$, $\tilde{\alpha}\!\restriction\! A = \alpha$, $\varphi \circ \tilde{g} = \tilde{\alpha} \cdot (\varphi\!\restriction\! U)$, $M(\log \circ \tilde{\alpha}) < 2^{-n-1}$, and $\tilde{g}(Y_{\tau(\kappa_i)} \cap U) = Y_{\tau(\kappa_i)} \cap U$ for $0 \leq i < l$. We put $h_{n+1}\!\restriction\! U = \tilde{g}$ and we note that hypotheses (3), (6), and (7) for $n+1$ and U are trivially satisfied.

We define for each $x \in f_n^{-1}(U)$,

(7.14) $\qquad\qquad\qquad \beta_{n+1}(x) = \tilde{\alpha}(f_n(x))\beta_n(x)$

and we note that hypothesis (8) is satisfied for $n+1$. Concerning hypothesis (9) for $n+1$ and U we have the following straightforward computation:

(7.15)
$$\begin{aligned}
(\beta_{n+1}\cdot\varphi) \circ f_{n+1}^{-1}\!\restriction\! U &= (\beta_{n+1}\cdot\varphi) \circ f_n^{-1} \circ \tilde{g}^{-1} \\
&= ((\tilde{\alpha} \circ f_n)\cdot\beta_n\cdot\varphi) \circ f_n^{-1} \circ \tilde{g}^{-1} \\
&= (\tilde{\alpha}\cdot((\beta_n\cdot\varphi) \circ f_n^{-1})) \circ \tilde{g}^{-1} \\
&= (\tilde{\alpha}\cdot\varphi) \circ \tilde{g}^{-1} \\
&= \varphi\!\restriction\! U,
\end{aligned}$$

where we used hypothesis (9) for n.

Still for Case II we now consider hypothesis (4). Let $m \in E(n+1)$ and note that by hypothesis $f_n(X_{\tau(m)}) \cap W = f_m(X_{\tau(m)}) \cap W$. We may assume that $f_n(X_{\tau(m)}) \cap U = f_m(X_{\tau(m)}) \cap U \neq \emptyset$. Because of the way $\mathcal{U}_{n+1}$ was constructed from $\mathcal{V}$ we have $f_n(X_{\tau(n+1)} \cap X_{\tau(m)}) \cap U \neq \emptyset$. In view of the Remark and the monotonicity of τ_E

we have $m = \nu_i$ for some $i < l$. Consequently,

(7.16) $f_{n+1}(X_{\tau(m)}) \cap U = h_{n+1}(f_n(X_{\tau(\nu_i)}) \cap U) = h_{n+1}(Y_{\tau(\kappa_i)} \cap U)$
$= Y_{\tau(\kappa_i)} \cap U = f_n(X_{\tau(m)}) \cap U = f_m(X_{\tau(m)}) \cap U.$

Finally, we verify hypothesis (5) for Case II. Let $k \in F(n+1)$ and note that by hypothesis $f_n^{-1}(Y_{\tau(k)} \cap W) = f_k^{-1}(Y_{\tau(k)} \cap W)$. We may assume that $Y_{\tau(k)} \cap U \neq \emptyset$. Because of the way $\mathcal{U}_{n+1}$ was constructed from $\mathcal{V}$ we have $f_n(X_{\tau(n+1)}) \cap Y_{\tau(k)} \cap U \neq \emptyset$ and hence $Y_{\tau(\kappa_{l-1})} \cap Y_{\tau(k)} \cap U \neq \emptyset$. We first consider the case $|\tau(k)| \leq l-1 = |\tau(\kappa_{l-1})|$. In view of the Remark we have $k = \kappa_i$ for some $i < l$. Consequently,

(7.17) $f_{n+1}^{-1}(Y_{\tau(k)} \cap U) = f_n^{-1}(h_{n+1}^{-1}(Y_{\tau(\kappa_i)} \cap U))$
$= f_n^{-1}(Y_{\tau(k)} \cap U)$
$= f_k^{-1}(Y_{\tau(k)} \cap U).$

Now, let $|\tau(k)| \geq l$. There exists a $j \in F(k+1) \subset F(n+1)$ such that $\tau(k){\restriction}l = \tau(j)$. By hypothesis (7) there is an $m \in E$ such that $Y_{\tau(j)} \cap U = f_n(X_{\tau(m)}) \cap U$ and $|\tau(j)| = |\tau(m)| = l$. Since

(7.18) $\emptyset \neq f_n(X_{\tau(n+1)}) \cap Y_{\tau(k)} \cap U$
$\subset f_n(X_{\tau(n+1)}) \cap Y_{\tau(j)} \cap U$
$= f_n(X_{\tau(n+1)} \cap X_{\tau(m)}) \cap U$

and $|\tau(n+1)| = l = |\tau(m)|$ we have by Remark 7.4 that $\tau(n+1) = \tau(m)$ and hence this situation is covered by Case I. The proof is complete. $\square$

REMARK 7.7. Note that in Theorem 7.5 if $t > |\log(M(\psi)/M(\varphi))|$ then we can as in Theorems 6.2 and 6.4 arrange that $M(\log \circ \beta) < t$.

DEFINITION 7.8. SL is the class of all bounded USC functions $\varphi\colon X \to \mathbb{R}^+$ such that X is a zero-dimensional space for which there exists a Sierpiński stratification $(X_s)_{s \in T}$ with the following properties:

(a) if $s \in T$ and $t \in \mathrm{succ}(s)$, then $G_0^{\varphi {\restriction} X_t}$ is nowhere dense in $G_0^{\varphi {\restriction} X_s}$ and
(b) if $s \in T$ then $\varphi {\restriction} X_s$ is a Lelek function.

We require that the elements of SL are bounded because that condition simplifies the following result.

LEMMA 7.9. If $\varphi \in \mathsf{SL}$ then there is a compactification C of the domain X of φ such that $(\mathrm{ext}_C \varphi, X) \in \mathsf{SLC}$.

PROOF. Let $\varphi \in \mathsf{SL}$ and let $(X_s)_{s \in T}$ be a system as in Definition 7.8. Let C be a zero-dimensional compactification of X such that for every X_s we have $(\mathrm{ext}_C \varphi) {\restriction} \overline{X}_s = \mathrm{ext}_{\overline{X}_s}(\varphi {\restriction} X_s)$ as in Lemma 4.8.c. Let $\psi = \mathrm{ext}_C \varphi$ and let $Y_s = \overline{X}_s$ for $s \in T$. Since φ is bounded we have that $\psi(X) \subset \mathbb{R}^+$. Note that conditions (1)–(3) of Definition 7.2 are trivially satisfied. Let $s \in T$. We have that $\bigcup\{G_0^{\varphi {\restriction} X_t} : t \in \mathrm{succ}(s)\} = G_0^{\varphi {\restriction} X_s}$ is a subset of $\bigcup\{G_0^{\psi {\restriction} Y_t} : t \in \mathrm{succ}(s)\}$ that is dense in $L_0^{\psi {\restriction} Y_s}$ by Lemma 4.8.b. So $(Y_s)_{s \in T}$ satisfies condition (5). For condition (4) note that if $t \in \mathrm{succ}(s)$ then since $\varphi {\restriction} X_t$ is Lelek we have that $G_0^{\varphi {\restriction} X_t}$ is dense in $G_0^{\psi {\restriction} Y_t}$, again by Lemma 4.8.b. Now $G_0^{\psi {\restriction} Y_t}$ is nowhere dense in $G_0^{\psi {\restriction} Y_s}$ because $G_0^{\varphi {\restriction} X_t}$ is nowhere dense in $G_0^{\varphi {\restriction} X_s}$. $\square$

Lemma 7.9 combines with Theorem 7.5 to:

THEOREM 7.10. *Any two elements φ and ψ of* SL *are m-equivalent (and hence G_0^φ is homeomorphic to G_0^ψ).*

PROPOSITION 7.11. *The function η (formula (2.4)) is an element of* SL.

PROOF. If we define $T = \mathbb{Q}^{<\omega}$ and $X_{q_0\ldots q_{k-1}} = \{q_0\} \times \cdots \times \{q_{k-1}\} \times \mathbb{Q} \times \mathbb{Q} \times \cdots$, then it is a straightforward exercise to show that $\eta : \mathbb{Q}^\omega \to \mathbb{R}^+$ is an element of SL. □

Noting that G_0^η is homeomorphic to $\mathfrak{E}$ we find that Proposition 7.11 combines with Theorems 7.5, 7.10, and Lemma 7.9 to prove the following characterization theorem.

THEOREM 7.12. *The following statements about a space E are equivalent:*
 (1) *E is homeomorphic to $\mathfrak{E}$,*
 (2) *there is a pair $(\varphi, X) \in$ SLC such that $G_0^{\varphi \restriction X}$ is homeomorphic to E,*
 (3) *there is a function $\varphi \in$ SL such that G_0^φ is homeomorphic to E,*

CHAPTER 8

Intrinsic characterizations of Erdős space

It is easily verified that it follows from condition (b) in Definition 7.8 that $\{x \in X \colon \varphi(x) = 0\}$ is a dense G_δ-set in X, alternatively, consider the prototype $\eta \in \mathsf{SL}$. This means that in order to fit a space into the characterization theorems of Chapter 7 we have to extend the space. Moreover, to use Theorem 7.12 we have to start with a very particular imbedding of the Erdős space candidate into $\mathfrak{C} \times \mathbb{R}$ respectively $\mathbb{Q}^\omega \times \mathbb{R}$, where $\mathfrak{C}$ is the Cantor set. These facts limit the power of the theorems in Chapter 7. We will now present a number of characterizations of Erdős space in terms of internal topological properties of the space. These intrinsic characterizations will turn out to be more powerful and easier to use than the results of the preceding chapter.

DEFINITION 8.1. Let T be a tree and let $(X_s)_{s \in T}$ be a system of subsets of a space X such that $X_t \subset X_s$ whenever $s \prec t$. A subset A of X is called an *anchor* for $(X_s)_{s \in T}$ in X if for every $\sigma \in [T]$ we have $X_{\sigma \restriction k} \cap A = \emptyset$ for some $k \in \omega$ or the sequence $X_{\sigma \restriction 0}, X_{\sigma \restriction 1}, \ldots$ converges to a point in X.

Thus the anchor A has the property that for every sequence that is generated by an element of $[T]$ if it is attached to A then it must converge and cannot be free to drift out of the space. Note that if $(X_s)_{s \in T}$ is a Sierpiński stratification, then the whole space is an anchor.

REMARK 8.2. Let Y be an $F_{\sigma\delta}$-space that is a witness to the almost zero-dimensionality of a space X. Thus X is a subset but not necessarily a subspace of Y and we let Z be the set X with the topology that is inherited from Y. Let $(Y_s)_{s \in T}$ be a Sierpiński stratification of Y and put $Z_s = Y_s \cap Z$ for $s \in T$. Let $x \in X$ and choose a neighbourhood B of x in X such that B is closed in Y. If $\sigma \in [T]$ is such that $Y_{\sigma \restriction k} \cap B \neq \emptyset$ for each $k \in \omega$ we have that $Y_{\sigma \restriction 0}, Y_{\sigma \restriction 1}, \ldots$ converges in Y to a point that must lie in B. Hence $Z_{\sigma \restriction 0}, Z_{\sigma \restriction 1}, \ldots$ converges in Z and we have that B is an anchor for $(Z_s)_{s \in T}$ in Z.

Consider now the special case that Y is an $F_{\sigma\delta}$-subset of $\mathbb{R}^\omega$ with $\dim Y = 0$. Let $p > 0$ and let $X = \ell^p \cap Y$ with the norm topology. Since the norm is LSC on $\mathbb{R}^\omega$ we have that every set of the form $\{x \in X \colon \|x\| \leq N\}$ is closed in Y. This means that every bounded set in X is an anchor for $(Z_s)_{s \in T}$ in Z.

DEFINITION 8.3. E is the class of all nonempty spaces E such that there exists a topology $\mathfrak{T}$ on E that witnesses the almost zero-dimensionality of E and there exist a nonempty tree T over a countable set and subspaces E_s of E that are closed with respect to $\mathfrak{T}$ for each $s \in T$ such that:

(1) $E_\emptyset = E$ and $E_s = \bigcup \{E_t \colon t \in \mathrm{succ}(s)\}$ whenever $s \in T$,
(2) each $x \in E$ has a neighbourhood U that is an anchor for $(E_s)_{s \in T}$ in $(E, \mathfrak{T})$,

39

(3) for each $s \in T$ and $t \in \mathrm{succ}(s)$ we have that E_t is nowhere dense in E_s, and
(4) E is $\{E_s \colon s \in T\}$-cohesive.

In view of Corollary 4.20 condition (4) is equivalent to:

($\hat{4}$) each $x \in E$ has a neighbourhood U such that for every $s \in T$ and every clopen set C in E if $C \cap E_s \subset U$ then $C \cap E_s = \emptyset$.

This condition may be slightly easier to verify than (4) since we only have to consider sets that are clopen in the whole space.

We will show that E is simply the class of all spaces that are homeomorphic to Erdős space.

REMARK 8.4. The anchor concept in Definition 8.3 is essential. This is because condition (4) excludes the possibility that the whole E is an anchor for $(E_s)_{s \in T}$, that is, we cannot have a Sierpiński stratification.

Aiming for a contradiction, let $\mathcal{T}$ be a topology that witnesses the almost zero-dimensionality of a space E and let $(E_s)_{s \in T}$ be a Sierpiński stratification of $(E, \mathcal{T})$ such that E is $\{E_s \colon s \in T\}$-cohesive. We can then find a countable collection $\mathcal{B} = \{B_i \colon i \in \omega\}$ such that

(1) for every $x \in X$ and every neighbourhood U of x there is a $B \in \mathcal{B}$ such that $x \in \mathrm{int}\,B \subset B \subset U$,
(2) every $B \in \mathcal{B}$ is closed with respect to $\mathcal{T}$, and
(3) every $B \in \mathcal{B}$ fails to contain nonempty clopen subsets of any E_s.

We may put $B_0 = \emptyset$. We now construct inductively a sequence $s_0 \prec s_1 \prec \cdots$ of nodes of T such that $E_{s_n} \neq \emptyset$, $B_n \cap E_{s_n} = \emptyset$ and $|s_n| \geq n$ for each $n \in \omega$. Put $s_0 = \emptyset$. Assume now that s_n has been found and consider B_{n+1}. Property (3) obviously implies that there is an $x \in E_{s_n} \setminus B_{n+1}$. Since we have a Sierpiński stratification there is a $\tau \in [T]$ such that $s_n \prec \tau$ and $E_{\tau \restriction 0}, E_{\tau \restriction 1}, \ldots$ converges to x in $(E, \mathcal{T})$. Since B_{n+1} is closed in $(E, \mathcal{T})$ we have that there is $k > |s_n|$ with $B_{n+1} \cap E_{\tau \restriction k} = \emptyset$. If we put $s_{n+1} = \tau \restriction k$ then the induction is finished. There obviously is a $\sigma \in [T]$ such that $s_n \prec \sigma$ for every n. Consequently, there is a point $y \in \bigcap_{k=0}^{\infty} E_{\sigma \restriction k} = \bigcap_{n=0}^{\infty} E_{s_n}$. Since there is an n with $y \in B_n$ we have a contradiction with the property $B_n \cap E_{s_n} = \emptyset$.

LEMMA 8.5. *If $E \in \mathsf{E}$ then there is an $\chi \in \mathsf{SL}$ such that E is homeomorphic to G_0^χ.*

PROOF. Let $E \in \mathsf{E}$ and let $\mathcal{T}$ and $(E_s)_{s \in T}$ be as in Definition 8.3. We first sketch the outline of the proof. We begin by taking the stratification $(E_s)_{s \in T}$ through a 'refining' process such that $Z = (E, \mathcal{T})$ admits a zero-dimensional extension X that also witnesses the almost zero-dimensionality of E and with the property that the closures $(X_r)_{r \in T'}$ of the sets in the refined system form a Sierpiński stratification of X, that is, the whole space becomes an anchor. With Lemma 4.11 we can find a USC function $\varphi : X \to \mathbb{I}$ such that $G_0^\varphi \approx E$. Condition (4) of Definition 8.3 allows us to replace φ by a function χ such that $G_0^\varphi \approx G_0^\chi$ and $\chi \restriction \overline{X}_r$ is a Lelek function for each $r \in T'$, see Lemma 5.9. We then have that $\chi \in \mathsf{SL}$.

First note that since Z is zero-dimensional we can arrange as in Lemma 7.3 that the system $(E_s)_{s \in T}$ satisfies the disjointness condition

(5) for all $s, t \in T$ if $|s| = |t|$ then $s = t$ or $E_s \cap E_t = \emptyset$.

By the fact that $\mathcal{T}$ is a witness topology and condition (2) we can find a countable collection $\mathcal{B}$ of subsets of E such that
  (a) for every $x \in E$ and every neighbourhood U of x there is a $B \in \mathcal{B}$ such that $x \in \operatorname{int} B \subset B \subset U$,
  (b) every $B \in \mathcal{B}$ is closed with respect to $\mathcal{T}$,
  (c) every $B \in \mathcal{B}$ is an anchor for $(E_s)_{s \in T}$ in Z.

We now construct by induction a sequence $\mathcal{C}_0 \subset \mathcal{C}_1 \subset \ldots$ of countable boolean algebras consisting of clopen subsets of the zero-dimensional space Z. Let $\mathcal{C}_0$ be a such a countable boolean algebra that is also a basis for the topology $\mathcal{T}$. Assume that $\mathcal{C}_n$ has been constructed. Let A_1, A_2 be two elements of the collection $\mathcal{A} = \mathcal{B} \cup \{E_s : s \in T\}$ and let $C_1, C_2 \in \mathcal{C}_n$. If $(A_1 \cap C_1) \cap (A_2 \cap C_2) = \emptyset$ select a clopen set $D = D(A_1, C_1, A_2, C_2)$ in Z such that $A_1 \cap C_1 \subset D$ and $D \cap A_2 \cap C_2 = \emptyset$. Otherwise, put $D(A_1, C_1, A_2, C_2) = Z$. We define $\mathcal{C}_{n+1}$ as the boolean algebra that is generated by

(8.1) $\qquad \mathcal{C}_n \cup \{D(A_1, C_1, A_2, C_2) : A_1, A_2 \in \mathcal{A} \text{ and } C_1, C_2 \in \mathcal{C}_n\}.$

Note that $\mathcal{C} = \bigcup_{n=0}^{\infty} \mathcal{C}_n$ is a countable boolean algebra consisting of clopen subsets of Z such that $\mathcal{C}$ is a basis for $\mathcal{T}$ and for every $A_1, A_2 \in \mathcal{A}$ and $C_1, C_2 \in \mathcal{C}$ with $(A_1 \cap C_1) \cap (A_2 \cap C_2) = \emptyset$ there is a $D \in \mathcal{C}$ such that $A_1 \cap C_1 \subset D$ and $D \cap A_2 \cap C_2 = \emptyset$.

Let $\{C_i : i \in \mathbb{N}\}$ be an enumeration of $\mathcal{C}$. We consider the tree $S = \{0,1\}^{<\omega}$ and we put

(8.2) $\qquad D_s = \bigcap \{C_i : i \leq l, s_i = 1\} \cap \bigcap \{E \setminus C_i : i \leq l, s_i = 0\},$

where $s = s_1 \ldots s_l \in S$ and $\bigcap \emptyset = E$. Note that $(D_s)_{s \in S}$ satisfies conditions (1) and (5) and that every D_s is an element of $\mathcal{C}$. Consider now the product tree $T * S$ and define $E'_{t*s} = E_t \cap D_s$ for each $t * s \in T * S$. Let $T' = \{r \in T * S : E'_r \neq \emptyset\}$.

Since every E'_{t*s} is the intersection of E_t with a set that is clopen in the topology $\mathcal{T}$ it follows easily that the system $(E'_r)_{r \in T'}$ satisfies the conditions of Definition 8.3. In fact, $(E'_r)_{r \in T'}$ satisfies condition (c). Also the system satisfies the disjointness condition (5) because the two contributing systems do. Moreover, we have that if $A_1, A_2 \in \mathcal{B} \cup \{E'_r : r \in T'\}$ with $A_1 \cap A_2 = \emptyset$, then $A_1 = A'_1 \cap C'_1$ and $A_2 = A'_2 \cap C'_2$ for some $A'_1, A'_2 \in \mathcal{A}$ and $C'_1, C'_2 \in \mathcal{C}$. Thus there is a $D \in \mathcal{C}$ that separates A_1 from A_2. Observe also that if $k \in \mathbb{N}$ and $r \in T'$ with $|r| \geq k$, then $E'_r \subset C_k$ or $C_k \cap E'_r = \emptyset$.

Let K be the Stone space that corresponds to the algebra $\mathcal{C}$. Thus K is a zero-dimensional compactification of Z. We let $\overline{A}$ denote the closure of A in K. Let Y_r be the closure of E'_r (seen as a subset of Z) in K. Note that if $\sigma \in [T']$ then $Y_{\sigma \restriction k} \neq \emptyset$ for every k and hence there is an $x_\sigma \in \bigcap_{k=0}^{\infty} Y_{\sigma \restriction k}$. Since basic neighbourhoods of x_σ are of the form $\overline{C_k}$ where $C_k \in \mathcal{C}$ and since $Y_{\sigma \restriction k} \subset \overline{C_k}$ we have that the sequence $Y_{\sigma \restriction 0}, Y_{\sigma \restriction 1}, \ldots$ converges to x_σ.

We put $X = \{x_\sigma : \sigma \in [T']\}$ and note that because of condition (1) we have $Z \subset X$. Let $X_r = Y_r \cap X$ for every $r \in T'$ and note that every X_r is nonempty because $E'_r \subset X_r$. We now verify that X is a witness to the almost zero-dimensionality of E. It suffices to prove that every $B \in \mathcal{B}$ is closed in X. Let $x_\sigma \in X$ and $B \in \mathcal{B}$. If there is a $k \in \omega$ such that $B \cap E'_{\sigma \restriction k} = \emptyset$, then we can find a $D \in \mathcal{C}$ with $B \subset D$ and $D \cap E'_{\sigma \restriction k} = \emptyset$. Thus the clopen set $\overline{D}$ separates B from $Y_{\sigma \restriction k}$ in K and hence $x_\sigma \notin \overline{B}$. If $E'_{\sigma \restriction k} \cap B \neq \emptyset$ for every k then $E'_{\sigma \restriction 0}, E'_{\sigma \restriction 1}, \ldots$ converges in the topology $\mathcal{T}$ to a point of E which has to lie in B because B is closed in the witness topology. This point has to be equal to x_σ (which is the limit of the closures of $E'_{\sigma\restriction 0}, E'_{\sigma\restriction 1}, \ldots$) so we have $x_\sigma \in B$.

Let $s, t \in T'$ such that $|s| = |t|$ and $s \neq t$. Then $E'_s \cap E'_t = \emptyset$ and hence there is a $D \in \mathcal{C}$ that separates the two sets. Thus Y_s and Y_t must also be disjoint. So we have that $(Y_r)_{r \in T'}$ satisfies the disjointness condition (5). We now verify that $(X_r)_{r \in T'}$ is a Sierpiński stratification for X. It is obvious that the system satisfies condition (ii), that $X_\emptyset = X$, and that $X_t \subset X_s$ whenever $s \prec t$. Let $s \in T'$ and let $x_\sigma \in X_s$. Put $k = |s|$ and note that $x_\sigma \in Y_s$ and $x_\sigma \in Y_{\sigma\restriction k}$. By disjointness we have that $\sigma\restriction k = s$. Put $t = \sigma\restriction(k+1)$ and note that $x_\sigma \in Y_t \cap X = X_t$ with $t \in \mathrm{succ}(s)$. So $(X_r)_{r \in T'}$ also satisfies condition (i) and is a Sierpiński stratification of X.

By Lemma 4.11 there exists a USC function $\varphi \colon X \to \mathbb{I}$ such that $h(x) = (x, \varphi(x))$ defines a homeomorphism from E to G_0^φ. Note that $h(E'_r) = G_0^{\varphi\restriction X_r}$ and that $\{x \in X_r : \varphi(x) > 0\} = E'_r$ is dense in X_r for each $r \in T'$. With condition (4) for the system $(E'_r)_{r \in T'}$ and Lemma 5.9 we find a USC function $\chi \colon X \to \mathbb{R}^+$ such that $\{x \in X : \chi(x) > 0\} = Z$, such that $g(x, \varphi(x)) = (x, \chi(x))$ for $x \in Z$ defines a homeomorphism from G_0^φ to G_0^χ, and such that every $\chi\restriction X_s$ is a Lelek function, which corresponds to condition (b) of Definition 7.8. Let $s, t \in T'$ be such that $t \in \mathrm{succ}(s)$. Note that $g(h(E'_r)) = G_0^{\chi\restriction X_r}$ for each $r \in T'$ and hence condition (3) of Definition 8.3 produces condition (a) of Definition 7.8. This completes the proof that $\chi \in \mathsf{SL}$ and we have $G_0^\chi \approx G_0^\varphi \approx E$. $\square$

Lemma 8.5 combines with Theorem 7.10 to:

THEOREM 8.6. *Any two elements of E are homeomorphic.*

DEFINITION 8.7. E' is the class of all nonempty spaces E such that there exists an $F_{\sigma\delta}$-topology $\mathcal{T}$ on E that witnesses the almost zero-dimensionality of E and there exist a nonempty tree T over a countable set and subspaces E_s of E that are closed with respect to $\mathcal{T}$ for each $s \in T \setminus \{\emptyset\}$ such that:

(1') $E_\emptyset$ is dense in E and $E_s = \bigcup \{E_t : t \in \mathrm{succ}(s)\}$ whenever $s \in T$,
(2') each $x \in E$ has a neighbourhood U that is an anchor for $(E_s)_{s \in T}$ in $(E, \mathcal{T})$,
(3') for each $s \in T \setminus \{\emptyset\}$ and $t \in \mathrm{succ}(s)$ we have that E_t is nowhere dense in E_s,
(4') E is $\{E_s : s \in T\}$-cohesive, and
(5') E can be written as a countable union of nowhere dense subsets that are closed with respect to $\mathcal{T}$.

THEOREM 8.8. $\mathsf{E} = \mathsf{E}'$.

PROOF. Comparing Definition 8.3 with Definition 8.7 we immediately see that $(1) \Rightarrow (1')$, $(2) \Rightarrow (2')$, $(3) \Rightarrow (3')$, $(4) \Rightarrow (4')$, and $(1)\&(3) \Rightarrow (5')$. In addition, it follows from condition (2) and the fact that $\mathcal{T}$ is a witness topology that every point has a neighbourhood U in E that is $\mathcal{T}$-closed such that U seen as a subspace of $(E, \mathcal{T})$ has a Sierpiński stratification. Consequently, $(E, \mathcal{T})$ is a countable union of closed sets that are absolute $F_{\sigma\delta}$ and hence $(E, \mathcal{T})$ is an absolute $F_{\sigma\delta}$-space. Thus $\mathsf{E} \subset \mathsf{E}'$.

We now prove that $\mathsf{E}' \subset \mathsf{E}$. Let $E \in \mathsf{E}'$ with associated topology $\mathcal{T}$ and system $(E_s)_{s \in T}$. Let ρ and d be metrics for E respectively $Z = (E, \mathcal{T})$ such that $\rho \geq d$. Let diam_ρ, diam_d, U_ε^ρ, and U_ε^d denote diameters and open ε-neighbourhoods. We choose a Sierpiński stratification $(Z_s)_{s \in S}$ of the space Z such that every Z_s is

8. INTRINSIC CHARACTERIZATIONS OF ERDŐS SPACE

nonempty. Assume that S is a tree over a countable set A such that $A \cap T = \emptyset$. With condition (5′) we may also assume that Z_a is nowhere dense in E for every $a \in A' = \{s \in S : |s| = 1\}$. Our proof now consists in carefully 'grafting' the stratification $(E_t)_{t \in T}$ onto $(Z_s)_{s \in S}$ so that the combined stratification $(X_r)_{r \in \mathfrak{T}}$ satisfies Definition 8.3.

Let $a \in A'$ be fixed. Select for every $s \in S$ with $a \prec s$ a countable dense subset of Z_s (with respect to ρ). Let D^a be the union of these dense sets so D^a is a countable set with the property that $D^a \cap Z_s$ is ρ-dense in Z_s whenever $a \prec s$. Let $\{p_i^a : i \in \mathbb{N}\}$ be an enumeration of D^a such that $\{j : p_j^a = p_i^a\}$ is infinite for each $i \in \mathbb{N}$. With condition (1′) we can select for every $x \in E_\emptyset$ a $\tau(x) \in [T]$ such that $x \in \bigcap_{k=0}^\infty E_{\tau(x) \restriction k}$. Condition (2′) implies that $E_{\tau(x) \restriction 0}, E_{\tau(x) \restriction 1}, \ldots$ converges to x in Z. We now select recursively a sequence of points $q_1^a, q_2^a, \ldots$ from $E_\emptyset$ and numbers $k_1^a, k_2^a, \ldots$ in $\mathbb{N} \setminus \{1\}$ so that for every $n \in \mathbb{N}$ we have, using the abbreviation $\theta_n^a = \tau(q_n^a) \restriction k_n^a$,

(a) $\rho(p_n^a, q_n^a) < 1/n$,
(b) $\mathrm{diam}_d E_{\theta_n^a} < 1/n$,
(c) $E_{\theta_n^a} \cap Z_a = \emptyset$, and
(d) $E_{\theta_n^a} \cap E_{\theta_i^a} = \emptyset$ for each $i < n$.

Assume that $n \in \mathbb{N}$ and that q_i^a and k_i^a have been found for $1 \leq i < n$. Since $F = Z_a \cup \bigcup_{i=1}^{n-1} E_{\theta_i^a}$ is nowhere dense in E and since $E_\emptyset$ is dense in E we can find a $q_n^a \in E_\emptyset \setminus F$ with $\rho(p_n^a, q_n^a) < 1/n$. Since F is closed in Z we can find a $k_n^a \in \mathbb{N} \setminus \{1\}$ such that $E_{\tau(q_n^a) \restriction k_n^a} \cap F = \emptyset$ and $\mathrm{diam}_d E_{\tau(q_n^a) \restriction k_n^a} < 1/n$. This completes the induction. Note that since $q_n^a \in E_{\theta_n^a}$ and $d \leq \rho$ we have that $E_{\theta_n^a} \subset U_{2/n}^d(p_n^a)$.

We will construct a stratification $(X_r)_{r \in \mathfrak{T}}$ that satisfies the conditions of Definition 8.3. $\mathfrak{T}$ will be a tree over $A \cup T$ that contains S. Begin by putting $X_\emptyset = E$. Let $s \in S \setminus \{\emptyset\}$ and $a = s \restriction 1$. We define

(8.3) $$N_s = \{n \in \mathbb{N} : n \geq |s| \text{ and } p_n^a \in Z_s\}$$

and

(8.4) $$X_s = Z_s \cup \bigcup \{E_{\tau(q_n^a) \restriction (k_n^a + |s|)} : n \in N_s\}.$$

Since Z_s and every E_t is closed in Z and since $E_{\tau(q_n^a) \restriction (k_n^a + |s|)} \subset E_{\theta_n^a} \subset U_{2/n}^d(p_n^a)$ we have that X_s is closed in Z and that $X_s \subset U_{2/|s|}^d(Z_s)$. We now define the following tree over $A \cup T$:

(8.5) $$\mathfrak{T} = S \cup \{s{\frown}t_1 \ldots t_l : s \in S \setminus \{\emptyset\}, a = s \restriction 1, l \in \mathbb{N}, n \in N_s,$$
$$t_0 = \tau(q_n^a) \restriction (k_n^a + |s|), \text{ and } t_i \in \mathrm{succ}(t_{i-1}) \text{ in } T \text{ for } 1 \leq i \leq l\}.$$

If $r = s{\frown}t_1 \ldots t_l \in \mathfrak{T} \setminus S$ then we define

(8.6) $$X_r = E_{t_l}.$$

It is left to verify that $(X_r)_{r \in \mathfrak{T}}$ satisfies conditions (1)–(4) of Definition 8.3.

Condition (1). $X_\emptyset = E$ by definition. Note that $E = \bigcup_{a \in A'} Z_a \subset \bigcup\{X_r : r \in \mathfrak{T}, |r| = 1\} \subset E$. Let $r, r' \in \mathfrak{T}$ with $|r| \geq 1$ and $r' \in \mathrm{succ}(r)$. Put $a = r \restriction 1 = r' \restriction 1$. If $r' \in S$ then $r \in S$ and $Z_{r'} \subset Z_r$. If $n \in N_{r'}$ then $n \geq |r'| > |r|$ and $n \in N_r$. Thus we have that $E_{\tau(q_n^a) \restriction (k_n^a + |r'|)} \subset E_{\tau(q_n^a) \restriction (k_n^a + |r|)}$ and hence $X_{r'} \subset X_r$. If $r' \in \mathfrak{T} \setminus S$ then $r' = s{\frown}t_1 \ldots t_l$ and $X_{r'} = E_{t_l}$ with $t_1 \in \mathrm{succ}(\tau(q_n^a) \restriction (k_n^a + |s|))$ for some $n \in N_s$ and $a = s \restriction 1$. If $l > 1$ then $r = s{\frown}t_1 \ldots t_{l-1}$ and $E_{t_l} \subset E_{t_{l-1}} = X_r$ because

$t_l \in \operatorname{succ}(t_{l-1})$. If $l = 1$ then $r = s$ and $X_{r'} = E_{t_1} \subset E_{\tau(q_n^a) \restriction (k_n^a + |r|)} \subset X_r$. Thus we have $\bigcup \{X_{r'} : r' \in \operatorname{succ}(r)\} \subset X_r$.

For the converse inclusion, let $r \in \mathfrak{T} \setminus \{\emptyset\}$ with $a = r \restriction 1$. First, let $r \in S$ and $X_r = Z_r \cup \bigcup \{E_{\tau(q_n^a) \restriction (k_n^a + |s|)} : n \in N_r\}$. We have $Z_r = \bigcup \{Z_s : s \in S \cap \operatorname{succ}(r)\} \subset \bigcup \{X_s : s \in S \cap \operatorname{succ}(r)\}$. Consider now an $x \in E_{\tau(q_n^a) \restriction (k_n^a + |s|)}$ with $n \in N_r$. Then there exists a $t \in \operatorname{succ}(\tau(q_n^a) \restriction (k_n^a + |s|))$ with $x \in E_t$. Note that $r' = r {\frown} t \in \operatorname{succ}(r) \subset \mathfrak{T}$ and that $X_{r'} = E_t$. Secondly, let $r \in \mathfrak{T} \setminus S$ and $x \in X_r$. Then by the same reasoning there is an $r' = r {\frown} t$ with $x \in X_{r'}$.

Condition (2). Let $x \in E$ and let U be a neighbourhood of x that is an anchor for $(X_t)_{t \in T}$ in Z. Let $\xi \in [\mathfrak{T}]$ such that $U \cap X_{\xi \restriction i} \neq \emptyset$ for all $i \in \omega$. If $\xi \in [S]$ then $Z_{\xi \restriction 0}, Z_{\xi \restriction 1}, \ldots$ converges to a point z in Z. Since $Z_{\xi \restriction i} \subset X_{\xi \restriction i} \subset U_{2/i}^d(Z_{\xi \restriction i})$ for each i we have that also $X_{\xi \restriction 0}, X_{\xi \restriction 1}, \ldots$ converges to z with respect to d. Consider now the case $\xi \in [\mathfrak{T}] \setminus [S]$. Then ξ must have the form $s {\frown} t_1 t_2 \ldots$ with $s \in S \setminus \{\emptyset\}$ and $t_{i+1} \in \operatorname{succ}(t_i) \subset T$ for every $i \in \mathbb{N}$. Let $\chi \in [T]$ be such that $t_i \prec \chi$ for each $i \in \mathbb{N}$. Put $l = |s|$ and note that $X_{\xi \restriction (l+i)} = E_{t_i}$ for each $i \in \mathbb{N}$ which implies that $U \cap E_{\chi \restriction j} \neq \emptyset$ for all $j \in \omega$. Since U is an anchor we have that $E_{\chi \restriction 0}, E_{\chi \restriction 1}, \ldots$ converges to a point y in Z. Since a tail of this sequence is identical to $X_{\xi \restriction (l+1)}, X_{\xi \restriction (l+2)}, \ldots$ we have that $X_{\xi \restriction 0}, X_{\xi \restriction 1}, \ldots$ converges also to y in Z.

Condition (3). Consider $r, r' \in \mathfrak{T}$ with $r' \in \operatorname{succ}(r)$. We first assume that $r' \in S$ (and hence also $r \in S$). Let $x \in X_{r'}$ and let $\varepsilon > 0$. Since $X_{r'}$ is closed in Z and hence in E it suffices to prove that there is a $z \in X_r \setminus X_{r'}$ with $\rho(x, z) < \varepsilon$. Since $r' \in S$ we have $X_{r'} = Z_{r'} \cup \bigcup \{E_{\tau(q_n^a) \restriction (k_n^a + |r'|)} : n \in N_{r'}\}$ with $a = r' \restriction 1$. We may assume that there is an $n \in N_{r'}$ and a $y \in E_{\tau(q_n^a) \restriction (k_n^a + |r'|)}$ such that $\rho(x, y) < \varepsilon/2$ as follows. If $x \in Z_{r'}$ then we select an $n \geq \max\{|r'|, 4/\varepsilon\}$ such that $p_n^a \in Z_{r'} \cap U_{\varepsilon/4}^\rho(x)$. Putting $y = q_n^a$ we find that $y \in E_{\tau(q_n^a) \restriction (k_n^a + |r'|)}$ and $\rho(x, y) < \varepsilon/2$ using property (a). The same argument that showed that $X_{r'}$ is closed also proves that

(8.7) $$F = Z_{r'} \cup \bigcup \{E_{\tau(q_i^a) \restriction (k_i^a + |r'|)} : i \in N_{r'} \setminus \{n\}\}$$

is a closed subset of Z that does not contain y by properties (c) and (d). Since $E_{\tau(q_n^a) \restriction (k_n^a + |r'|)}$ is by condition (3′) nowhere dense in $E_{\tau(q_n^a) \restriction (k_n^a + |r'| - 1)}$ we have that there exists a $z \in E_{\tau(q_n^a) \restriction (k_n^a + |r'| - 1)} \setminus E_{\tau(q_n^a) \restriction (k_n^a + |r'|)}$ with $\rho(y, z) < \varepsilon/2$ and $z \notin F$. So $z \notin X_{r'}$ and $\rho(x, z) < \varepsilon$. If $r = \emptyset$ then $z \in E = X_r$ and we are finished. Let $r \neq \emptyset$ and hence $a = r \restriction 1$. We have $|r| < |r'| \leq n$ and $Z_{r'} \subset Z_r$ thus $n \in N_r$. Consequently,

(8.8) $$z \in E_{\tau(q_n^a) \restriction (k_n^a + |r'| - 1)} = E_{\tau(q_n^a) \restriction (k_n^a + |r|)} \subset X_r.$$

If $r' \in \mathfrak{T} \setminus S$ then $r' = s {\frown} t_1 \ldots t_l$ and $X_{r'} = E_{t_l}$ with t_1 an immediate successor of $t_0 = \tau(q_n^a) \restriction (k_n^a + |s|)$ for some $n \in N_s$ and $a = s \restriction 1$. Note that $X_{r'} = E_{t_l}$ is nowhere dense in $E_{t_{l-1}}$ because $t_l \in \operatorname{succ}(t_{l-1})$ and $t_{l-1} \neq \emptyset$. If $l > 1$ then $r = s {\frown} t_1 \ldots t_{l-1}$ and $X_r = E_{t_{l-1}}$. If $l = 1$ then $r = s$ and $E_{t_0} \subset X_r$.

Condition (4). Let $x \in E$ and let U be a neighbourhood of x such that U contains no nonempty clopen subsets of any E_t with the ρ-topology. Let C be a nonempty clopen subset of some X_r with the ρ-topology that is contained in U. If $r \in \mathfrak{T} \setminus S$ then $X_r = E_t$ for some $t \in T$ so we have that $r \in S$. Since $X_\emptyset = \bigcup\{X_s : |s| = 1\}$ we may assume that $r \neq \emptyset$ and we may put $a = r \restriction 1$. If $C \not\subset Z_r$ then C meets some $E_{\tau(q_n^a) \restriction (k_n^a + |r|)}$ that is contained in X_r so we may conclude that $C \subset Z_r$. Since $D^a \cap Z_r$ is dense in Z_r and C is clopen in X_r, both with respect to the ρ-topology, we may select a $p_i^a \in C \cap Z_r$. Because of the way D^a was enumerated we may choose a $j \in \mathbb{N}$ with $p_j^a = p_i^a$, $j \geq |r|$, and $X_r \cap U_{1/j}^\rho(p_j^a) \subset C$. Then $j \in N_r$ and $q_j^a \in U_{1/j}^\rho(p_j^a)$. Since $q_j^a \in E_{\tau(q_j^a) \restriction (k_j^a + |r|)} \subset X_r$ we have that $q_j^a \in C \cap E_{\tau(q_j^a) \restriction (k_j^a + |r|)}$, a contradiction. □

DEFINITION 8.9. For the next three results consider a fixed sequence $E_0, E_1, E_2, \ldots$ of subsets of $\mathbb{R}$ and let
$$\mathcal{E} = \{z \in \ell^p : z_n \in E_n \text{ for every } n \in \omega\}$$
be a corresponding subspace of some fixed ℓ^p.

The following two results were proved in Dijkstra [**16**].

THEOREM 8.10. *If $\mathcal{E}$ is not empty and every E_n is zero-dimensional, then the following statements are equivalent:*
   (1) *there exists an $x \in \prod_{n=0}^\infty E_n$ with $\|x\| = \infty$ and $\lim_{n \to \infty} x_n = 0$,*
   (2) *every nonempty clopen subset of $\mathcal{E}$ is unbounded,*
   (3) *$\mathcal{E}$ is cohesive, and*
   (4) *$\dim \mathcal{E} > 0$.*

Recall that if $A_0, A_1, \ldots$ is a sequence of subsets of a space X, then $\limsup_{n \to \infty} A_n = \bigcap_{n=0}^\infty \overline{\bigcup_{k=n}^\infty A_k}$.

COROLLARY 8.11. *If 0 is a cluster point of $\limsup_{n \to \infty} E_n$, then every nonempty clopen subset of $\mathcal{E}$ is unbounded (and hence $\dim \mathcal{E} \neq 0$).*

We now show that Theorems 8.6 and 8.8 are not void.

PROPOSITION 8.12. *Let $\dim \mathcal{E} > 0$ and let every E_n be an $F_{\sigma\delta}$-subset of $\mathbb{R}$ that is zero-dimensional. If infinitely many of the E_n's are of the first category in themselves, then $\mathcal{E} \in \mathsf{E}$ and $\mathcal{E}$ is homeomorphic to $\mathfrak{E}$.*

PROOF. We begin by re-ordering the E_n's such that E_n is of the first category in itself for every even n. Recall that the p-norm $\|\cdot\|$ is an LSC function from $\mathbb{R}^\omega$ to $[0, \infty]$. We let X be the (zero-dimensional) product space $\prod_{n=0}^\infty E_n \subset \mathbb{R}^\omega$ and we note that since the norm is LSC on X we have that X witnesses the almost zero-dimensionality of $\mathcal{E}$. Let $\mathcal{T}$ be the witness topology on $\mathcal{E}$ that is inherited from X. Since E_n is an $F_{\sigma\delta}$-space we may choose a Sierpiński stratification $(Z_s^n)_{s \in T_n}$ for E_n such that every $Z_s^n \neq \emptyset$. Since E_{2n} is of the first category in itself we may assume that for every $t \in T_{2n}$ with $|t| = 1$ we have that Z_t^{2n} is nowhere dense in $Z_\emptyset^{2n} = E_{2n}$. We now construct a tree $\mathfrak{T}$ as follows:

(8.9) $\mathfrak{T} = \{(s_0, \ldots, s_k, s_0', \ldots, s_k') : s_i \in T_{2i}, s_i' \in T_{2i+1},$
$$\text{and } |s_i| = |s_i'| = k - i \text{ for } 0 \leq i \leq k \text{ where } k \in \omega\},$$

where if $s = (s_0, \ldots, s_k, s_0', \ldots, s_k') \in \mathfrak{T}$ and $t = (t_0, \ldots, t_l, t_0', \ldots, t_l') \in \mathfrak{T}$, then $s \prec t$ means that $k \leq l$, $s_i \prec t_i$ and $s_i' \prec t_i'$ for every $i \leq k$. Observe that although $\mathfrak{T}$ does not formally satisfy Definition 3.4 it is obviously isomorphic to a countable tree. Note also that in this interpretation we have $|(s_0, \ldots, s_k, s_0', \ldots, s_k')| = k$. Let $s = (s_0, \ldots, s_k, s_0' \ldots, s_k') \in \mathfrak{T}$ and define the following closed subset of X:

(8.10) $\quad X_s = \{x \in X : x_{2i} \in Z_{s_i}^{2i} \text{ and } x_{2i+1} \in Z_{s_i'}^{2i+1} \text{ for } i \leq k\}$.

Let $\mathcal{E}_s$ stand for $\ell^p \cap X_s$ with the norm topology. Since $\dim \mathcal{E} > 0$ we have that statement (1) of Theorem 8.10 is valid for $\mathcal{E}$. Note that (1) remains valid if we replace a finite number of the E_n's by other nonempty sets thus it follows that every nonempty, clopen subset of $\mathcal{E}_s$ is unbounded which means that we find that $\mathcal{E}$ is $\{\mathcal{E}_s : s \in \mathfrak{T}\}$-cohesive just by choosing bounded neighbourhoods for the points of $\mathcal{E}$.

It is easily verified that $(X_s)_{s \in \mathfrak{T}}$ is a Sierpiński stratification of X because it is a product of Sierpiński stratifications. This means that also $(\mathcal{E}_s)_{s \in \mathfrak{T}}$ satisfies condition (1) of Definition 8.3 and that since X is a witness condition (2) is easily seen to be satisfied as well, see Remark 8.2. We now verify condition (3) of Definition 8.3. Let $s = (s_0, \ldots, s_k, s'_0, \ldots, s'_k) \in \mathfrak{T}$ and let $t = (t_0, \ldots, t_{k+1}, t'_0, \ldots, t'_{k+1}) \in \mathrm{succ}(s)$. Let $\varepsilon > 0$ and let $x \in \mathcal{E}_t$. Then $x_{2k} \in Z^{2k}_{t_k}$. Since $|t_k| = (k+1) - k = 1$ we have that $Z^{2k}_{t_k}$ is nowhere dense in E_{2k}. Select a $q \in E_{2k} \setminus Z^{2k}_{t_k}$ with $|q - x_{2k}| < \varepsilon$ and define $y \in \ell^p \setminus \mathcal{E}_t$ by

$$(8.11) \qquad y_i = \begin{cases} q, & \text{if } i = 2k; \\ x_i, & \text{if } i \neq 2k. \end{cases}$$

We have $\|y - x\| = |q - x_{2k}| < \varepsilon$. Since $x_{2i} \in Z^{2i}_{t_i} \subset Z^{2i}_{s_i}$ for $i < k$, $q \in E_{2k} = Z^{2k}_{\emptyset} = Z^{2k}_{s_k}$, and $x_{2i+1} \in Z^{2i+1}_{t'_i} \subset Z^{2i+1}_{s'_i}$ for $i \leq k$ we have that $y \in \mathcal{E}_s$. This completes the proof that $\mathcal{E} \in \mathsf{E}$. Note that if we define $\varphi(x) = 1/(1 + \|x\|)$ on X then it is easily seen that $\varphi \in \mathsf{SL}$. In particular, η (formula (2.4)) is in SL.

It is obvious that $\mathfrak{E}$ is one of the spaces that satisfies the conditions so the homeomorphy of $\mathcal{E}$ and $\mathfrak{E}$ follows from Theorem 8.6. □

Proposition 8.12 combines with Theorems 8.6 and 8.8 to prove the following characterization theorem.

THEOREM 8.13. *The following statements about a space E are equivalent:*
(1) *E is homeomorphic to $\mathfrak{E}$,*
(2) *$E \in \mathsf{E}$, and*
(3) *$E \in \mathsf{E}'$.*

REMARK 8.14. At first glance there does not appear to be much difference between Definitions 8.3 and 8.7. This, however, is a false impression. To use Definition 8.3 to prove that a given space E is homeomorphic to $\mathfrak{E}$ we have to construct a stratification of the entire space whereas condition (1′) of Definition 8.7 requires only a stratification of a dense subset of E. Let us examine the consequences if the Erdős space candidate E is for instance a topological group. Then we need only three things to satisfy Definition 8.7: an $F_{\sigma\delta}$ witness topology that has the property that group translations are homeomorphisms, the first category property (5′), and a suitable closed imbedding of Erdős space in E. Because if we have a copy $\mathcal{E}$ of $\mathfrak{E}$ in E of the right type which means in particular that it is also a closed imbedding on the level of the respective witness topologies, then we can obtain the dense stratified set $E_\emptyset$ by simply multiplying $\mathcal{E}$ with a countable dense subset of the group E. In effect, the condition $E \in \mathsf{E}'$ can be proved by using universality type argument similar to those used in zero-dimensional and infinite-dimensional topology.

This is the method that we will use to classify homeomorphism groups in Chapter 10. The particular imbeddings of Erdős space that we will use come from Dijkstra and van Mill [21] and Dijkstra [15] where we constructed them for the purpose of showing that the homeomorphism groups in question are one-dimensional.

COROLLARY 8.15. *If $O \subset \mathfrak{E}$ is nonempty and open, then O is homeomorphic to $\mathfrak{E}$.*

PROOF. Let $\mathfrak{T}$ and $(E_s)_{s \in T}$ be a witness topology respectively a stratification for $\mathfrak{E}$ as in Definition 8.3. If O is a nonempty open subset of $\mathfrak{E}$, then we define $\mathfrak{T}' = \{A \cap O \colon A \in \mathfrak{T}\}$, $E'_s = E_s \cap O$ for $s \in T$, and $T' = \{s \in T \colon E'_s \ne \emptyset\}$. Clearly, $\mathfrak{T}'$ is a witness topology for O, every E'_s is closed in $(O, \mathfrak{T}')$, and T' is tree. It is obvious that $(E'_s)_{s \in T'}$ satisfies conditions (1) and (3) of Definition 8.3 and that condition (4) follows from Remark 5.2. For condition (2) choose for each point in O a neighbourhood $U \subset O$ that is closed with respect to $\mathfrak{T}$ and that satisfies condition (2) for the system $(E_s)_{s \in T}$. If $\sigma \in [T]$ is such that $U \cap E'_{\sigma \restriction k} = U \cap E_{\sigma \restriction k} \ne \emptyset$ for each $k \in \omega$, then $E_{\sigma \restriction 0}, E_{\sigma \restriction 1}, \ldots$ converges in $(\mathfrak{E}, \mathfrak{T})$ to a point x that must lie in U (and hence in O) because U is closed with respect to $\mathfrak{T}$. Then $E'_{\sigma \restriction 0}, E'_{\sigma \restriction 1}, \ldots$ converges also to x in $(O, \mathfrak{T}')$. Thus $O \in \mathsf{E}$. □

Kawamura, Oversteegen, and Tymchatyn [**31**] proved that Corollary 8.15 is also valid for complete Erdős space.

LEMMA 8.16. *We may replace condition (4) in Definition 8.3 and condition (4′) in Definition 8.7 by the following weaker condition:*

(4*) *If x is a point and U is a neighbourhood of x in E, then there is a neighbourhood V of x in E such that whenever an E_s meets V then it also meets $U \setminus V$.*

PROOF. (4) ⇒ (4*). Assume that E satisfies (4) and let U be an open neighbourhood of some point x in E. Select a neighbourhood W of x in E such that W contains no non-empty clopen subsets of any E_s. Select a neighbourhood V of x in E such that $V \subset U \cap W$ and V is closed in $(E, \mathfrak{T})$. Suppose that $E_s \cap V \ne \emptyset$ and $E_s \cap U \setminus V = \emptyset$. Note that $C = V \cap E_s$ is $\mathfrak{T}$-closed and therefore also closed in E_s. On the other hand, $C = U \cap E_s$ is $\mathfrak{T}$-open in E_s. Thus we have that V and hence W contain a nonempty clopen subset C of E_s. Since this contradicts the cohesion assumption we have proved property (4*).

(1)&(2)&(4*) ⇒ (4). Assume (1), (2), and (4*) and let x be a point in E. Let U be a neighbourhood of x in E that is an anchor for $(E_s)_{s \in T}$ in $(E, \mathfrak{T})$. Suppose that C is a nonempty clopen subset of some E_s that is contained in U. With property (4*) choose for each $x \in C$ a neighbourhood $V(x)$ of x in E such that $V(x) \subset C \cup (E \setminus E_s)$, $V(x)$ is closed in $(E, \mathfrak{T})$, and $E_t \cap (C \cup (E \setminus E_s)) \setminus V(x) \ne \emptyset$ whenever $E_t \cap V(x) \ne \emptyset$. In particular, we have that $E_t \cap V(x) \ne \emptyset$ implies $E_t \cap C \setminus V(x) \ne \emptyset$ whenever $s \prec t$. Since E is separable metric we can find a countable set $\{a_i : i \in \mathbb{N}\} \subset C$ with $C = E_s \cap \bigcup \{V(a_i) : i \in \mathbb{N}\}$. Since $E_s \setminus C$ is open in E_s we can use Remark 2.5 to write $E_s \setminus C = \bigcup_{i=1}^{\infty} F_i$ where every F_i is closed in $(E, \mathfrak{T})$. We now construct recursively a sequence $t_0 \precnsim t_1 \precnsim \cdots$ in T such that for every $i \in \omega$,

(a) $E_{t_i} \cap C \ne \emptyset$ and
(b) $E_{t_i} \cap \bigcup_{j=1}^{i}(V(a_j) \cup F_j) = \emptyset$.

We put $t_0 = s$ and note that the induction hypotheses are trivially satisfied. Assume that t_i has been found. If $E_{t_i} \cap V(a_{i+1}) \ne \emptyset$ then we have $E_{t_i} \cap C \setminus V(a_{i+1}) \ne \emptyset$ and if $E_{t_i} \cap V(a_{i+1}) = \emptyset$ then we also have $E_{t_i} \cap C \setminus V(a_{i+1}) \ne \emptyset$ because $E_{t_i} \cap C \ne \emptyset$. Let $x \in E_{t_i} \cap C \setminus V(a_{i+1})$ and select a $\sigma \in [T]$ such that $t_i \prec \sigma$ and $x \in \bigcap_{k=1}^{\infty} E_{\sigma \restriction k}$. Note that x is outside of the $\mathfrak{T}$-closed set $V(a_{i+1}) \cup F_{i+1}$. Since x is an element of the anchor U the sequence $E_{\sigma\restriction 0}, E_{\sigma\restriction 1}, \ldots$ converges to x in $(E, \mathfrak{T})$ and we may select a $k > |t_i|$ such that $E_{\sigma\restriction k} \cap (V(a_{i+1}) \cup F_{i+1}) = \emptyset$. Put $t_{i+1} = \sigma\restriction k$ and note that $x \in E_{t_{i+1}} \cap C$. Also observe that $t_i \subsetneqq t_{i+1}$ and $E_{t_{i+1}} \subset E_{t_i}$ thus hypothesis (b) is satisfied for $i+1$. The induction is complete.

Since $t_0 \subsetneqq t_1 \subsetneqq \cdots$ we can find a $\tau \in [T]$ with $\tau\restriction|t_i| = t_i$ for every $i \in \omega$. By property (a) and the fact that C is contained in the anchor U there is a $y \in E$ such that $y \in \bigcap_{k=0}^{\infty} E_{\tau\restriction k} = \bigcap_{i=0}^{\infty} E_{t_i}$. By property (b) we have that y is no element of $\bigcup_{j=1}^{\infty}(V(a_j) \cup F_j)$, which is a set that contains E_s. Since $y \in E_{t_0} = E_s$ we have the contradiction that proves condition (4).

Note that condition (4*) is strictly weaker than (4): it is easily seen that for instance the space $\mathbb{Q}$ admits a $\mathfrak{T}$ and $(E_s)_{s \in T}$ that satisfy (1), (3), and (4*), but obviously not (4). □

Using condition (4*) we can now formulate the following characterization theorems that correspond to E respectively E'.

THEOREM 8.17. *A nonempty space E is homeomorphic to $\mathfrak{E}$ if and only if there exists a zero-dimensional topology $\mathfrak{T}$ on E that is coarser than the given topology on E and there exist a nonempty tree T over a countable set and subspaces E_s of E that are closed with respect to $\mathfrak{T}$ for each $s \in T$ such that:*

(1) $E_\emptyset = E$ and $E_s = \bigcup\{E_t : t \in \mathrm{succ}(s)\}$ *whenever* $s \in T$,
(2) *for each $s \in T$ and $t \in \mathrm{succ}(s)$ we have that E_t is nowhere dense in E_s, and*
(3) *if x is a point and U is a neighbourhood of x in E, then there is a neighbourhood $V \subset U$ of x in E that is a closed anchor for $(E_s)_{s \in T}$ in $(E, \mathfrak{T})$ with the property that whenever an E_s meets V then it also meets $U \setminus V$.*

THEOREM 8.18. *A nonempty space E is homeomorphic to $\mathfrak{E}'$ if and only if there exists a zero-dimensional $F_{\sigma\delta}$-topology $\mathfrak{T}$ on E that is coarser than the given topology on E and there exist a nonempty tree T over a countable set and subspaces E_s of E that are closed with respect to $\mathfrak{T}$ for each $s \in T \setminus \{\emptyset\}$ such that:*

(1) $E_\emptyset$ *is dense in E and* $E_s = \bigcup\{E_t : t \in \mathrm{succ}(s)\}$ *whenever* $s \in T$,
(2) *for each $s \in T \setminus \{\emptyset\}$ and $t \in \mathrm{succ}(s)$ we have that E_t is nowhere dense in E_s,*
(3) E *can be written as a countable union of nowhere dense subsets that are closed with respect to $\mathfrak{T}$, and*
(4) *if x is a point and U is a neighbourhood of x in E, then there is a neighbourhood $V \subset U$ of x in E that is a closed anchor for $(E_s)_{s \in T}$ in $(E, \mathfrak{T})$ with the property that whenever an E_s meets V then it also meets $U \setminus V$.*

Proposition 8.26 and Theorem 8.27 of the preprint version are now Proposition 8.12 and Theorem 8.13.

CHAPTER 9

Factoring Erdős space

We begin by noting an interesting connection between Erdős space and complete Erdős space.

PROPOSITION 9.1. $\mathfrak{E}_c \times \mathbb{Q}^\omega$ *is homeomorphic to* $\mathfrak{E}$.

PROOF. Consider the sequence $E_0, E_1, \ldots$ of subsets of $\mathbb{R}$ that is defined by $E_{2n} = \mathbb{Q} \cap (-2^{-n}, 2^{-n})$ and $E_{2n+1} = \{0\} \cup \{1/m : m \in \mathbb{N}\}$ for $n \in \omega$. Let $\mathcal{E}$ be defined as in Definition 8.9 with $p = 2$. It is easily seen that Corollary 8.11 and Proposition 8.12 apply so $\mathcal{E}$ is homeomorphic to $\mathfrak{E}$. We obviously have that $\mathcal{E}$ is homeomorphic to the product of $\mathfrak{E}_c$ and

(9.1) $\qquad Z = \{x \in \ell^2 : x_n \in \mathbb{Q} \cap (-2^{-n}, 2^{-n}) \text{ for each } n \in \omega\}.$

Since it is well-known that the norm topology on Z coincides with the topology of coordinate-wise convergence we have that Z is homeomorphic to $\mathbb{Q}^\omega$. □

This proposition implies that the product of every zero-dimensional $F_{\sigma\delta}$-space with $\mathfrak{E}$ is homeomorphic to $\mathfrak{E}$, see van Engelen [**28**, Theorem 4.5.2]. We improve on this result as follows. We call a space X an *Erdős space factor* if there is a space Y such that $X \times Y$ is homeomorphic to $\mathfrak{E}$.

THEOREM 9.2 (Stability). *For a nonempty space E the following statements are equivalent:*
 (1) $E \times \mathfrak{E}$ *is homeomorphic to* $\mathfrak{E}$,
 (2) E *is an Erdős space factor,*
 (3) E *is homeomorphic to a retract of* $\mathfrak{E}$,
 (4) E *admits an imbedding as a C-set in* $\mathfrak{E}$,
 (5) E *admits a closed imbedding into* $\mathfrak{E}$,
 (6) E *is homeomorphic to a G_δ-subset of* $\mathfrak{E}$, *and*
 (7) E *is almost zero-dimensional as witnessed by an $F_{\sigma\delta}$-topology.*

PROOF. (1) ⇒ (2), (2) ⇒ (3), (4) ⇒ (5), and (5) ⇒ (6) are trivial and (3) ⇒ (4) by (the easy half of) Theorem 4.19.

(6) ⇒ (7). Assume that E is a G_δ-subset of $\mathfrak{E}$. Consider the product topology on $\mathfrak{E}$ that is inherited from $\mathbb{Q}^\omega$ and recall that this topology witnesses the almost zero-dimensionality of $\mathfrak{E}$. Since the Hilbert norm is LSC with respect to the product topology we have that $\mathfrak{E}$ is an F_σ-subset of the $F_{\sigma\delta}$-space $\mathbb{Q}^\omega$. So the product topology on $\mathfrak{E}$ is an $F_{\sigma\delta}$-topology. Since this topology is a witness to almost zero-dimensionality and E is G_δ in $\mathfrak{E}$ we have that E is an $F_{\sigma\delta}$-set with respect to the product topology, see Remark 2.5. So the product topology is a witness to the almost zero-dimensionality of E and it is absolute $F_{\sigma\delta}$.

(7) ⇒ (1). Assume now that (7) is valid. Let Z be the space E equipped with the $F_{\sigma\delta}$-topology. Choose a Sierpiński stratification $(Z_t)_{t \in T}$ for Z. Choose a

witness topology and a system $(Y_s)_{s \in S}$ for the space $\mathfrak{E}$ that satisfies the conditions in Definition 8.3. Let Y be $\mathfrak{E}$ equipped with that witness topology. Consider the product tree $S * T$ and put $X = Y \times Z$ and $X_{s*t} = Y_s \times Z_t$ for $s * t \in S * T$. Clearly, X witnesses the almost zero-dimensionality of $\mathfrak{E} \times E$. By Remark 5.2 we have that the system $(X_r)_{r \in S*T}$ satisfies condition (4) of Definition 8.3. The other conditions are trivially satisfied so $\mathfrak{E} \times E \in \mathsf{E}$. Thus $\mathfrak{E} \times E \approx \mathfrak{E}$ by Theorem 8.13. □

Since $\mathfrak{E}$ is a universal space for the class of almost zero-dimensional spaces, Theorem 4.15, we have:

COROLLARY 9.3. *Every complete, almost zero-dimensional, nonempty space X is an Erdős space factor.*

If p is a point in a space X then the *weak product* $W(X,p)$ is defined by

(9.2) $\qquad W(X,p) = \{x \in X^\omega : x_i = p \text{ for all but finitely many } i \in \omega\}.$

COROLLARY 9.4. $\mathfrak{E}^\omega$ *and* $W(\mathfrak{E}, \mathbf{0})$ *are homeomorphic to* $\mathfrak{E}$.

PROOF. Let Z stand for the set $\mathfrak{E}$ equipped with an $F_{\sigma\delta}$-topology that witnesses the almost zero-dimensionality of $\mathfrak{E}$. Then the topologies on Z^ω and $W(Z, \mathbf{0})$ are witnesses to the almost zero-dimensionality of $\mathfrak{E}^\omega$ respectively $W(\mathfrak{E}, \mathbf{0})$. The product Z^ω is trivially an $F_{\sigma\delta}$-space and $W(Z, \mathbf{0})$ is a countable union of closed $F_{\sigma\delta}$-spaces so also absolutely $F_{\sigma\delta}$. Thus we have by Theorem 9.2 that $\mathfrak{E} \approx \mathfrak{E} \times \mathfrak{E}^\omega \approx \mathfrak{E}^\omega$ and $\mathfrak{E} \approx \mathfrak{E} \times W(\mathfrak{E}, \mathbf{0}) \approx W(\mathfrak{E}, \mathbf{0})$. □

COROLLARY 9.5. $(\mathfrak{E}_c \times \mathbb{Q})^\omega$ *is homeomorphic to* $\mathfrak{E}$.

PROOF. By Proposition 9.1 and Corollary 9.4 we have $(\mathfrak{E}_c \times \mathbb{Q})^\omega \approx (\mathfrak{E}_c \times \mathbb{Q}^\omega)^\omega \approx \mathfrak{E}^\omega \approx \mathfrak{E}$. □

REMARK 9.6. Interestingly, we now have that $\mathfrak{E}_c$ and $\mathfrak{E}_c^\omega$, which are nonhomeomorphic according to Dijkstra, van Mill, and Steprāns [23], stabilize to the same space $\mathfrak{E}$ when multiplied by the zero-dimensional space $\mathbb{Q}^\omega$. Stability theorems (and characterizations) for the spaces $\mathfrak{E}_c$ and $\mathfrak{E}_c^\omega$ can be found in Dijkstra and van Mill [22] and Dijkstra [19], respectively.

Let $\mathcal{T}$ be a witness topology on an almost zero-dimensional space X. According to Remark 4.12 X is an $F_{\sigma\delta}$-space whenever $(X, \mathcal{T})$ is an $F_{\sigma\delta}$-space and $(X, \mathcal{T})$ is a $G_{\delta\sigma\delta}$-space whenever X is an $F_{\sigma\delta}$-space. In view of Theorem 9.2 the following question is a natural one.

QUESTION 9.7. Are the Erdős space factors precisely the nonempty almost zero-dimensional $F_{\sigma\delta}$-spaces or, equivalently, does every almost zero-dimensional $F_{\sigma\delta}$-space admit an $F_{\sigma\delta}$ witness topology?

CHAPTER 10

Groups of homeomorphisms

If X is a topological space then $\mathcal{H}(X)$ denotes the group of autohomeomorphisms of X and if $A \subset X$ then $\mathcal{H}(X,A)$ stands for the subgroup $\{h \in \mathcal{H}(X) : h(A) = A\}$. We denote the identity element of $\mathcal{H}(X)$ by e_X. If O is an open subset of X then $\mathcal{H}_O(X) = \{h \in \mathcal{H}(X) \colon h{\restriction}(X \setminus O) = e_{X \setminus O}\}$ and $\mathcal{H}_O(X, A) = \mathcal{H}_O(X) \cap \mathcal{H}(X, A)$.

If X is compact then the choice of a topology for $\mathcal{H}(X)$ is straightforward: the compact-open topology coincides with the topology of uniform convergence with respect to any compatible metric for X and makes $\mathcal{H}(X)$ into a topological group that is a Polish space. If $A, B \subset X$ then we define $[A, B] = \{h \in \mathcal{H}(X) : h(A) \subset B\}$. Thus a subbasis for the topology on $\mathcal{H}(X)$ consists of the sets $[K, O]$, where K is compact and O is open in X. Note that the topology of point-wise convergence is in general neither metrizable nor compatible with the group structure.

For noncompact spaces the situation is more complex. In that case, the topology of uniform convergence depends on the metric that one chooses for X and it is usually much stronger than the compact-open topology. However, for locally compact X a natural choice for a separable metric topology is available: the topology that $\mathcal{H}(X)$ inherits from $\mathcal{H}(\alpha X)$, where αX is the one-point compactification. Since $\mathcal{H}(X) = \mathcal{H}_X(\alpha X)$ it is also a topological group and a Polish space. Note that the compact-open topology may, even for locally compact spaces, not be compatible with the group structure, in particular with the inverse operation. However, if every point in X has a neighbourhood that is a continuum, then the topology that is inherited from $\mathcal{H}(\alpha X)$ coincides with the compact-open topology, see Dijkstra [17] and Arens [6]. The case that $\mathcal{H}(X)$ is equipped with the compact-open topology for noncompact X is discussed separately in Remark 10.7.

If X is locally compact and $A \subset X$, then we think of $\mathcal{H}(X, A)$ as a subspace of $\mathcal{H}(X)$. So $\mathcal{H}(X, A)$ is a topological group and hence a homogeneous space. If D is a zero-dimensional dense subset of X, then according to Dijkstra and van Mill [21] the space $\mathcal{H}(X, D)$ is almost zero-dimensional. We are here interested in the case that D is a countable dense subset of X. Then the topology on $\mathcal{H}(X, D)$ that is generated by the subbasis $\{[\{d\}, O] : d \in D$ and O open in $X\}$ is called the *topology of pointwise convergence on D*. This topology $\mathcal{T}$ coincides with the topology that $\mathcal{H}(X, D)$ inherits from the zero-dimensional product space D^D via the injection $h \mapsto h{\restriction}D$ of $\mathcal{H}(X, D)$ into D^D. The topology $\mathcal{T}$ is in general not compatible with the group structure but if $f \in \mathcal{H}(X, D)$, then the map $h \mapsto h \circ f$ is a homeomorphism of $(\mathcal{H}(X, D), \mathcal{T})$.

THEOREM 10.1. *Let M be a compact space, let O an open subset of M, and let D_1 be a countable dense subset of O. If D_2 is a countable dense subset of $M \setminus O$, then the topology of pointwise convergence on $D_1 \cup D_2$ is an $F_{\sigma\delta}$-topology that witnesses* the almost zero-dimensionality of $\mathcal{H}_O(M,D_1)$ and hence $\mathcal{H}_O(M,D_1)$ is an Erdős space factor.

PROOF. Put $D = D_1 \cup D_2$ and let $\mathcal{T}$ be the (zero-dimensional) topology that $\mathcal{H}_O(M, D_1) = \mathcal{H}_O(M, D)$ inherits from D^D. Note that in order to prove that $\mathcal{T}$ witnesses the almost zero-dimensionality of $\mathcal{H}_O(M,D)$ it suffices to construct a neighbourhood subbasis for the identity $e = e_M$ consisting of sets that are closed with respect to $\mathcal{T}$ because multiplication with an $f \in \mathcal{H}_O(M, D)$ is a homeomorphism with respect to $\mathcal{T}$. Let $\mathcal{S}$ consist of all sets $[K, F] \cap \mathcal{H}_O(M, D)$ where K and F are closed subsets of M such that $K = \overline{\operatorname{int} K}$ and $K \subset \operatorname{int} F$. It is easily verified that $\mathcal{S}$ is a neighbourhood subbasis at e in $\mathcal{H}_O(M,D)$. Let $[K,F] \cap \mathcal{H}_O(M,D)$ be an arbitrary element of $\mathcal{S}$ and let $h \in \mathcal{H}_O(M,D) \setminus [K,F]$. Then $\{x \in K : h(x) \notin F\}$ is an open nonempty subset of K. Since $K = \overline{\operatorname{int} K}$ and D is dense we have that $D \cap K$ is dense in K and hence there is an $a \in D \cap K$ such that $h(a) \notin F$. Observe that $[\{a\}, M \setminus F] \cap \mathcal{H}_O(M, D)$ is an element of $\mathcal{T}$ that is disjoint from $[K, F]$.

We now verify that $\mathcal{T}$ is an absolute $F_{\sigma\delta}$-topology and hence that $\mathcal{H}_O(M,D)$ is an Erdős space factor by the Stability Theorem 9.2. Let $\{(A_i, B_i): i \in \mathbb{N}\}$ be a countable collection of pairs of open subsets of M with disjoint closures such that for every pair (A, B) of disjoint closed subsets of M there exists an $i \in \mathbb{N}$ with $A \subset A_i$ and $B \subset B_i$. Let ρ be a metric on M. Since D is countable the product D^D is an $F_{\sigma\delta}$-space. We define the following $F_{\sigma\delta}$-subsets of D^D:

$$(10.1) \qquad S = \bigcap_{a \in D_1} \bigcup_{b \in D_1} \{h \in D^D : h(b) = a\} \cap \bigcap_{c \in D_2} \{h \in D^D : h(c) = c\},$$

$$(10.2) \qquad F = \bigcap_{i=1}^{\infty} \bigcup_{n=1}^{\infty} \bigcap_{a \in A_i \cap D} \bigcap_{b \in B_i \cap D} \{h \in D^D : \rho(h(a), h(b)) \geq 1/n\},$$

$$(10.3) \qquad C = \bigcap_{i=1}^{\infty} \bigcup_{n=1}^{\infty} \bigcap_{\substack{a,b \in D \\ \rho(a,b) < 1/n}} \{h \in D^D : h(a) \notin A_i \text{ or } h(b) \notin B_i\}.$$

Note that S consists of all surjective elements of D^D that restrict to the identity on D_2. It suffices to show that the set $H = \{h {\restriction} D : h \in \mathcal{H}_O(M,D)\}$ coincides with $S \cap F \cap C$. It is easily verified that $H \subset S \cap F \cap C$.

Let $h \in S \cap F \cap C$. Let A and B be disjoint closed subsets of M and select an $i \in \mathbb{N}$ such that $A \subset A_i$ and $B \subset B_i$. Since $h \in C$ there is an $n \in \mathbb{N}$ such that $\rho(h^{-1}(A_i), h^{-1}(B_i)) \geq 1/n$. Consequently, we have $\rho(h^{-1}(A), h^{-1}(B)) \geq 1/n$ and hence $\overline{h^{-1}(A)} \cap \overline{h^{-1}(B)} = \emptyset$. According to [37, Lemma A.8.3] this means that h can be extended to a continuous $\overline{h}: M \to M$. Since $h \in S$ we have $h(D) = \overline{h}(D) = D$ and $h{\restriction}D_2 = \operatorname{id}_{D_2}$. Since D and D_2 are dense in M respectively $M \setminus O$ we have that $\overline{h}$ is a surjection that is supported on O. Let x and y be distinct points of M. Select an i such that $x \in A_i$ and $y \in B_i$. Since $h \in F$ there is an n such that $\rho(h(A_i \cap D), h(B_i \cap D)) \geq 1/n$ and hence $\rho(\overline{h(A_i \cap D)}, \overline{h(B_i \cap D)}) \geq 1/n$. Since A_i and B_i are open, D is dense, and $\overline{h}$ is continuous we have $\rho(\overline{h}(x), \overline{h}(y)) \geq 1/n$. So we may conclude that $\overline{h}$ is injective and hence $\overline{h} \in \mathcal{H}_O(M,D)$ and $h \in H$. □

A space X is called *strongly locally homogeneous* if the space has a basis $\mathcal{B}$ such that for every $B \in \mathcal{B}$ and $x, y \in B$ there is an autohomeomorphism h of X that is supported on B and that maps x to y. The spaces $\mathbb{R}^n$ and the Hilbert cube are well-known examples of such spaces. If a complete space X is strongly locally homogeneous, then for every open O in X and countable dense subsets D_1 and D_2 of O there is an autohomeomorphism h of X that is supported on O and that maps D_1 precisely onto D_2, see Bennett [7].

THEOREM 10.2. *Let M be a locally compact space, let O be an open subset of M, and let D be a countable dense subset of O. If O contains an open set that is a topological n-manifold with $n \geq 2$ or a Hilbert cube manifold, then $\mathcal{H}_O(M, D)$ is homeomorphic to Erdős space.*

PROOF. We use the method outlined in Remark 8.14. If M is not compact then $\mathcal{H}_O(M, D) = \mathcal{H}_O(\alpha M, D)$ so we may assume that M is compact. Let ρ be a metric on M and let $\hat{\rho}$ be the induced metric on $\mathcal{H}(M)$: $\hat{\rho}(f, g) = \max_{x \in M} \rho(f(x), g(x))$ for $f, g \in \mathcal{H}(M)$. Note that $\hat{\rho}$ is right invariant: $\hat{\rho}(f \circ h, g \circ h) = \hat{\rho}(f, g)$. We prove the theorem by showing that $\mathcal{H}_O(M, D) \in \mathsf{E}'$. Let D' be the union of D with a countable dense subset of $M \setminus O$ and let $\mathcal{T}$ be the topology that $\mathcal{H}_O(M, D) = \mathcal{H}_O(M, D')$ inherits from $D'^{D'}$. Thus according to Theorem 10.1 $\mathcal{T}$ is an $F_{\sigma\delta}$-topology that witnesses the almost zero-dimensionality of $\mathcal{H}_O(M, D)$. We let $\hat{\mathbb{R}}$ stand for the compactification $[-\infty, \infty]$ of $\mathbb{R}$. We shall use the convention $\pm\infty + t = \pm\infty$ when $t \in \mathbb{R}$, thereby extending addition to a continuous operation from $\hat{\mathbb{R}} \times \mathbb{R}$ to $\hat{\mathbb{R}}$. Let J stand for the interval $[-1, 1]$.

Consider first the case that O contains an open copy of $\mathbb{R}^n$ for some $n \geq 2$. We may then assume that O contains the n-cell $\hat{\mathbb{R}} \times J^{n-1}$ such that $\mathbb{R} \times (-1, 1)^{n-1}$ is the interior of $\hat{\mathbb{R}} \times J^{n-1}$ in M. By strong local homogeneity we may also assume that $D \cap (\mathbb{R} \times (-1, 1)^{n-1})$ equals the set $\mathbb{Q} \times Q$, where $Q = (\mathbb{Q} \cap (-1, 1))^{n-1} \setminus \{\theta\}$ and $\theta = (0, \ldots, 0) \in (-1, 1)^{n-1}$.

We define the Erdős space

(10.4) $$\mathcal{E} = \{z \in \ell^1 : z_i \in \mathbb{Q}^+ \text{ for every } i \in \omega\},$$

where $\mathbb{Q}^+ = \mathbb{Q} \cap [0, \infty)$. Note that for $z = (z_0, z_1, \ldots) \in \mathcal{E}$ we have $\|z\| = \sum_{i=0}^{\infty} z_i$. We will imbed $\mathcal{E}$ in $\mathcal{H}_O(M, D)$. For every $z \in \mathcal{E}$ we define the function $\alpha_z \colon \mathbb{I} \to \mathbb{R}$ by

(10.5) $$\alpha_z(r) = \begin{cases} z_i 2^{i+1}(2^{-i} - r) + \sum_{k=0}^{i-1} z_k, & \text{if } 2^{-i-1} \leq r \leq 2^{-i} \text{ for } i \in \omega; \\ \|z\|, & \text{if } r = 0. \end{cases}$$

Note that $\alpha_z(2^{-i}) = \sum_{k=0}^{i-1} z_k$ for $z \in \mathcal{E}$ and $i \in \omega$ and that α_z simply connects these points with linear segments. It is clear that α_z is well-defined and continuous and that

(10.6) $$|\alpha_z(r) - \alpha_{z'}(r)| \leq \|z - z'\|$$

for every $z, z' \in \mathcal{E}$ and $r \in \mathbb{I}$. Furthermore, we have that $\alpha_z(r) \in \mathbb{Q}$ whenever $r \in \mathbb{Q} \cap (0, 1]$.

If $y = (y_1, \ldots, y_{n-1}) \in J^{n-1}$ then we put $|y| = \max\{|y_1|, \ldots, |y_{n-1}|\}$. Note that if $y \in Q$ then $|y| \in \mathbb{Q} \cap (0, 1]$ and $\alpha_z(|y|) \in \mathbb{Q}$ for any $z \in \mathcal{E}$. For each $z \in \mathcal{E}$ we define the map $H_z \colon \hat{\mathbb{R}} \times J^{n-1} \to \hat{\mathbb{R}} \times J^{n-1}$ by

(10.7) $$H_z(x, y) = (x + \alpha_z(|y|), y).$$

Since α_z is continuous and bounded (by $\|z\|$) it is clear that H_z is well-defined and an element of $\mathcal{H}(\hat{\mathbb{R}} \times J^{n-1})$. Since $\alpha_z(1) = 0$ and $\pm\infty + \alpha_z(r) = \pm\infty$ we have that every H_z is supported on $\mathbb{R} \times (-1, 1)^{n-1}$ and hence we may assume that every H_z

has been extended with the identity to an element of $\mathcal{H}_O(M)$. Observe that for $(x,y) \in \mathbb{R} \times J^{n-1}$,

$$(10.8) \qquad d(H_z(x,y), H_{z'}(x,y)) = |\alpha_z(|y|) - \alpha_{z'}(|y|)| \leq \|z - z'\|,$$

where d is the standard euclidean metric that $\mathbb{R} \times J^{n-1}$ inherits from $\mathbb{R}^n$. This means that $H_z \restriction (\mathbb{R} \times J^{n-1})$ depends continuously on z if we use the topology of uniform convergence on $\mathcal{H}(\mathbb{R} \times J^{n-1})$ with respect to d. So we certainly have that $H \colon \mathcal{E} \to \mathcal{H}_O(M)$ is continuous. For $i \in \omega$ let $p_i = (2^{-i}, 0, \ldots, 0)$. Observe that for each $z \in \mathcal{E}$, $i \in \omega$, and $r \in \mathbb{R}$,

$$(10.9) \qquad H_z(r, \theta) = (r + \|z\|, \theta) \quad \text{and} \quad H_z(r, p_i) = \left(r + \sum_{k=0}^{i-1} z_k, p_i\right)$$

and hence $\pi(H_z(0, \theta)) = \|z\|$ and $\pi(H_z(0, p_{i+1})) - \pi(H_z(0, p_i)) = z_i$, where $\pi \colon \mathbb{R} \times J^{n-1} \to \mathbb{R}$ is the projection. This means that H is a one-to-one map. Since the norm topology is the weakest topology on ℓ^1 that makes the coordinate projections and the norm function continuous it follows that if we use H to pull the topology of point-wise convergence on $\mathcal{H}(M)$ back to $\mathcal{E}$, then we get a topology that is at least as strong as the norm topology. Since we already know that H is continuous and since the topology of uniform convergence is stronger than the topology of point-wise convergence we have that H is an imbedding. If $(x,y) \in \mathbb{Q} \times Q$ and $z \in \mathcal{E}$, then $\alpha_z(|y|) \in \mathbb{Q}$ so $H_z(x,y) = (x + \alpha_z(|y|), y) \in \mathbb{Q} \times Q$. If, on the other hand, $H_z(x,y) \in \mathbb{Q} \times Q$ then $y \in Q$ and thus $x \in \mathbb{Q} - \alpha_z(|y|) = \mathbb{Q}$. So H is an imbedding of $\mathcal{E}$ in $\mathcal{H}_O(M, D)$.

Consider the point $(0, p_1) \in \mathbb{Q} \times Q \subset D$. For every $a \in D$ we define $Y_a = \{h \in \mathcal{H}_O(M, D) \colon h(0, p_1) = a\}$ and we note that every Y_a is closed with respect to $\mathcal{T}$ and that $\bigcup_{a \in D} Y_a = \mathcal{H}_O(M, D)$. Let for $i \in \mathbb{N}$, $z^i = (1/i, 0, 0, \ldots) \in \mathcal{E}$ and note that $\lim_{i \to \infty} z^i = \mathbf{0}$ thus $\lim_{i \to \infty} H_{z^i} = H_\mathbf{0} = e_M$ in $\mathcal{H}_O(M, D)$. If $h \in Y_a$ then $\lim_{i \to \infty} h \circ H_{z^i} = h$ but $h \circ H_{z^i} \notin Y_a$ because $h(H_{z^i}(0, p_1)) = h(1/i, p_1) \neq h(0, p_1) = a$. Thus Y_a is nowhere dense in $\mathcal{H}_O(M, D)$ and condition $(5')$ of Definition 8.7 is satisfied.

Let Z denote $\mathcal{E}$ equipped with the witness topology that is inherited from $(\mathbb{Q}^+)^\omega$. We now verify that $H \colon Z \to (\mathcal{H}_O(M, D), \mathcal{T})$ is a closed imbedding. First we investigate continuity: if (x, y) is a fixed element of $\mathbb{Q} \times Q$ then there is an $i \in \omega$ such that $2^{-i-1} \leq |y| \leq 2^{-i}$. Note that $\alpha_z(|y|)$ is a linear function of $z_0, \ldots, z_i$ so also $H_z(x, y)$ depends continuously on only finitely many coordinates z_j. If $a \in D' \setminus (\mathbb{Q} \times Q)$ then $H_z(a) = a$ for all $z \in \mathcal{E}$. Thus $H_z(a)$ depends continuously on z for each $a \in D'$ which means that $H \colon Z \to (\mathcal{H}_O(M, D), \mathcal{T})$ is continuous. Now let $h \in \mathcal{H}_O(M, D)$ be such that there is a sequence $z^1, z^2, \ldots$ in Z with $\lim_{j \to \infty} H_{z^j}(a) = h(a)$ for every $a \in D'$. Since $\mathbb{R} \times (-1, 1)^{n-1}$ is the interior of $\hat{\mathbb{R}} \times J^{n-1}$ in M the set $A = D' \setminus (\mathbb{R} \times (-1, 1)^{n-1})$ is dense in $M \setminus (\mathbb{R} \times (-1, 1)^{n-1})$. Since every H_z is supported on $\mathbb{R} \times (-1, 1)^{n-1}$ the same is true for h. Thus $h(\mathbb{Q} \times Q) = \mathbb{Q} \times Q$, $h(0, p_0) = (0, p_0)$, and

$$(10.10) \qquad \begin{aligned} z_i &= \pi(h(0, p_{i+1})) - \pi(h(0, p_i)) \\ &= \lim_{j \to \infty} \left(\pi(H_{z^j}(0, p_{i+1})) - \pi(H_{z^j}(0, p_i))\right) \\ &= \lim_{j \to \infty} z_i^j \end{aligned}$$

is well-defined and an element of $\mathbb{Q}^+$ for each $i \in \omega$. By the definition of the z_k's we have $\pi(h(0, p_i)) = \sum_{k=0}^{i-1} z_k$. Since h is supported on $\mathbb{R} \times (-1,1)^{n-1}$ we have $\pi(h(0, \theta)) < \infty$ and hence

$$(10.11) \qquad \|z\| = \lim_{i \to \infty} \pi(h(0, p_i)) = \pi(h(\lim_{i \to \infty}(0, p_i))) = \pi(h(0, \theta)) < \infty.$$

So $z = (z_0, z_1, \dots) \in Z$ and $\lim_{j \to \infty} z^j = z$ in Z. Thus $h = H_z$ and we have that $H: Z \to (\mathcal{H}_O(M, D), \mathcal{T})$ is a closed imbedding.

Let $\pm\Omega = (\pm\infty, \theta) \in \hat{\mathbb{R}} \times J^{n-1}$. We now make an observation which will be the key to satisfying conditions $(2')$ and $(4')$ of the definition of E': if A is an unbounded subset of $\mathcal{E}$ then

$$(10.12) \qquad \mathrm{diam}_{\hat\rho}\{H_z \colon z \in A\} \geq \rho(-\Omega, \Omega).$$

Let $z \in A$ and let $n \in \mathbb{N}$ be arbitrary. Select a z^n in A such that $\|z^n\| > \|z\| + 2n$ and let $r = -\|z\| - n$. We have $H_z(r, \theta) = (-n, \theta)$ and $H_{z^n}(r, \theta) = (\|z^n\| - \|z\| - n, \theta)$. Hence,

$$(10.13) \qquad \begin{aligned} \mathrm{diam}_{\hat\rho}\{H_z \colon z \in A\} &\geq \limsup_{n \to \infty} \hat\rho(H_z, H_{z^n}) \\ &\geq \lim_{n \to \infty} \rho((-n, \theta), (\|z^n\| - \|z\| - n, \theta)) \\ &= \rho(-\Omega, \Omega). \end{aligned}$$

We now consider the natural stratification of $\mathcal{E}$ that satisfies Definition 8.3, cf. Proposition 8.12. Let $T = (\mathbb{Q}^+)^{<\omega}$ and for each $s = q_1 \dots q_k \in T$ we put

$$(10.14) \qquad \mathcal{E}_s = \{z \in \mathcal{E} \colon z_{i-1} = q_i \text{ for } 1 \leq i \leq k\}.$$

Note that according to Remark 8.2 every bounded subset of $\mathcal{E}$ is an anchor for $(\mathcal{E}_s)_{s \in T}$ in Z and that according to Corollary 8.11 every nonempty clopen subset of any $\mathcal{E}_s$ is unbounded. Let $F = \{f_q \colon q \in \mathbb{Q}^+\}$ be a countable dense subset of $\mathcal{H}_O(M, D)$. Since $H: Z \to (\mathcal{H}_O(M, D), \mathcal{T})$ is a closed map we have that $X_s = \{H_z \colon z \in \mathcal{E}_s\}$ is closed with respect to $\mathcal{T}$ for each $s \in T$. We define $(E_s)_{s \in T}$ as follows:

$$(10.15) \qquad E_\emptyset = X_\emptyset \circ F$$

and if $s = q_0 \dots q_k \in T \setminus \{\emptyset\}$ then

$$(10.16) \qquad E_s = X_{q_1 \dots q_k} \circ f_{q_0}.$$

Note that if $f \in \mathcal{H}_O(M, D)$ then the map $h \mapsto h \circ f$ is a homeomorphism of $(\mathcal{H}_O(M, D), \mathcal{T})$ as well as of $\mathcal{H}_O(M, D)$. So every E_s is closed with respect to $\mathcal{T}$ provided $s \neq \emptyset$.

All that remains is to show that $(E_s)_{s \in T}$ satisfies conditions $(1')$–$(4')$ of Definition 8.7. Since $H_\mathbf{0} = e_M$ we have that $E_\emptyset$ contains F and is dense. The other part of condition $(1')$ is equally trivial. Since $H: \mathcal{E} \to \mathcal{H}_O(M, D)$ is an imbedding we have that condition $(3')$ is satisfied. Now let U be an arbitrary set in $\mathcal{H}_O(M, D)$ such that $\mathrm{diam}_{\hat\rho} U < \rho(-\Omega, \Omega)$. We shall see that U works for condition $(2')$ as well as $(4')$. Let $\sigma = q_0 q_1 \dots \in [T]$ be such that $E_{\sigma \restriction k} \cap U \neq \emptyset$ for each $k \in \omega$. Putting $\tau = q_1 q_2 \dots \in [T]$ we have that $X_{\tau \restriction k} \cap (U \circ f_{q_0}^{-1}) \neq \emptyset$ for each $k \in \omega$. Since $\hat\rho$ is right invariant we have $\mathrm{diam}_{\hat\rho}(U \circ f_{q_0}^{-1}) < \rho(-\Omega, \Omega)$ and hence $V = \{z \in \mathcal{E} \colon H_z \in U \circ f_{q_0}^{-1}\}$ is bounded. Thus V is an anchor for $(\mathcal{E}_s)_{s \in T}$ in Z and obviously $\mathcal{E}_{\tau \restriction k} \cap V \neq \emptyset$ for each $k \in \omega$. Thus $\mathcal{E}_{\tau \restriction 0}, \mathcal{E}_{\tau \restriction 1}, \dots$ converges to an element z in Z. Then $X_{\tau \restriction 0}, X_{\tau \restriction 1}, \dots$ converges to H_z and $E_{\sigma \restriction 0}, E_{\sigma \restriction 1}, \dots$ converges to $H_z \circ f_{q_0}$, both with respect to $\mathcal{T}$. Thus condition $(2')$ is satisfied. Now let C be a nonempty clopen subset of some E_s such that $C \subset U$. We may assume that $|s| \geq 1$ and we put $q = s{\restriction}1$ and $q{\frown}t = s$. So $\text{diam}_{\hat\rho}(C \circ f_q^{-1}) < \rho(-\Omega, \Omega)$ and $C \circ f_q^{-1}$ is a nonempty clopen subset of X_t. This means that $\{z \in \mathcal{E}: H_z \in C \circ f_q^{-1}\}$ is a nonempty, clopen, bounded subset of $\mathcal{E}_t$. As mentioned above, this contradicts Corollary 8.11 so we may conclude that condition $(4')$ is satisfied and $\mathcal{H}_O(M, D) \in \mathsf{E}'$.

Consider now the case that O contains an open subset that is a $\mathfrak{Q}$-manifold, where $\mathfrak{Q} = \mathbb{I}^\omega$. Then we may assume that O contains the Hilbert cube $\hat{\mathbb{R}} \times J^{n-1} \times \mathfrak{Q}$ such that $\mathbb{R} \times (-1, 1)^{n-1} \times \mathfrak{Q}$ is an open subset of $\hat{\mathbb{R}} \times J^{n-1} \times \mathfrak{Q}$ in M, where $n \geq 2$. We may also assume that $D \cap (\mathbb{R} \times (-1, 1)^{n-1} \times \mathfrak{Q})$ equals the set $\mathbb{Q} \times Q \times C$, where C is some countable dense subset of $\mathfrak{Q}$. If we then replace H_z as defined above by $H_z \times \text{id}_\mathfrak{Q}$, then the proof for the Hilbert cube case is completely analogous the proof given above. $\square$

For $n \in \mathbb{N}$ let μ^n denote the universal Menger continuum of dimension n, see Engelking [**27**, §1.11] or Bestvina [**9**]. A nonempty space M is called a *Menger manifold* if there is an $n \in \mathbb{N}$ such that M has an open cover consisting of sets that are homeomorphic to open subsets of μ^n. According to [**9**, Theorem 3.2.2] every Menger manifold is strongly locally homogeneous.

LEMMA 10.3. *Let $f: X \to Y$ and $g: Y \to Z$ be continuous. If $g \circ f$ is a closed imbedding then so is f.*

PROOF. Since $g \circ f$ is one-to-one so is f. Let F be a closed subset of X. Since $g \circ f$ is closed and g is continuous we have that $A = g^{-1}((g \circ f)(F))$ is closed in Y. Note that $f \circ (g \circ f)^{-1} \circ g{\restriction}A$ is a retraction from A onto $f(F)$. Thus $f(F)$ is closed in A and Y. $\square$

THEOREM 10.4. *Let M be a locally compact space, let O be an open subset of M, and let D be a countable dense subset of O. If O contains an open set that is a Menger manifold, then $\mathcal{H}_O(M, D)$ is homeomorphic to Erdős space.*

PROOF. The beginning is identical to the first paragraph of the proof of Theorem 10.2. Let $n \in \mathbb{N}$ be such that O contains an nonempty open subset U that is homeomorphic to an open subset of μ^n. Select a null sequence $V_0, V_1, \ldots$ of nonempty, open sets such that their closures in M are disjoint subsets of U. Put $V = \bigcup_{k=0}^\infty V_k$. Consider the following complete Erdős space:

(10.17) $$E_3 = \{z \in \ell^1: 3^i z_i \in \omega \text{ for } i \in \omega\}$$

and let Z_3 stand for E_3 equipped with the witness topology that is inherited from the product $\mathbb{R}^\omega$. (The fact that E_3 is homeomorphic to $\mathfrak{E}_c$ follows from Dijkstra [**16**, Theorem 3] but is not used here.) If $i \in \omega$ then we let $\xi_i: E_3 \to E_3$ denote the projection that is given by the rule $\xi_i(z) = (z_0, z_1, \ldots, z_i, 0, 0, \ldots)$. We let P be the countable dense subset $\bigcup_{i=0}^\infty \xi_i(E_3)$ of E_3. Let $k \in \omega$. According to Dijkstra [**15**, Remarks 7 and 8] there exist a closed imbedding $G^k: E_3 \ni z \mapsto G_z^k \in \mathcal{H}_{V_k}(M)$, a copy $\hat{\mathbb{R}}_k$ of $\hat{\mathbb{R}}$ in V_k, and a sequence $p_1^k, p_2^k, \ldots \in V_k \setminus \hat{\mathbb{R}}_k$ such that

(a) $\lim_{i \to \infty} p_i^k = 0_k \in \mathbb{R}_k$, where $\mathbb{R}_k = \hat{\mathbb{R}}_k \setminus \{\pm\infty_k\}$,
(b) for each $r \in \hat{\mathbb{R}}_k$ and $z \in E_3$ we have $G_z^k(r) = r + \|z\| \in \hat{\mathbb{R}}_k$,
(c) for each $x \in M \setminus \mathbb{R}_k$ there is an $i \in \omega$ such that $G_z^k(x) = G_{\xi_i(z)}^k(x)$ for each $z \in E_3$, and

(d) $\beta_k \circ G^k \colon Z_3 \to \beta_k(\mathcal{H}(M))$ is a closed imbedding, where $A_k = \{\infty_k, p_1^k, p_2^k, \ldots\}$ and $\beta_k \colon \mathcal{H}(M) \to M^{A_k}$ is given by the rule $\beta_k(h) = h{\restriction}A_k$.

Consider the Cantor set

(10.18) $$C' = \{z \in E_3 \colon z_i \in \{0, 3^{-i}\} \text{ for } i \in \omega\}$$

and note that since $\sum_{i=0}^{\infty} 3^{-i} < \infty$ we have that on C' the norm topology coincides with the product topology. Let $\delta \colon C' \to \mathbb{R}^+$ be the imbedding that is given by the rule $\delta(z) = \|z\|$. We define $C = \delta(C')$, $\gamma = \delta^{-1}{\restriction}C$, and $Q = \delta(C' \cap P)$. Thus C is a Cantor set with Q as a countable dense subset and $\|\gamma(r)\| = r$ for each $r \in C$. We define a complete Erdős space

(10.19) $$\mathcal{E}_c = \{z \in \ell^1 \colon z_i \in C \text{ for } i \in \omega\}$$

and an Erdős space

(10.20) $$\mathcal{E} = \{z \in \ell^1 \colon z_i \in Q \text{ for } i \in \omega\}.$$

We let Z_c and Z stand for $\mathcal{E}_c$ respectively $\mathcal{E}$ with the witness topologies that these spaces inherit from $\mathbb{R}^\omega$. Let $\nu \colon \omega \times \omega \to \omega$ be a bijection such that $\nu(i,j) \geq j$ for all $i,j \in \omega$. We define an imbedding $\zeta \colon \mathcal{E}_c \to E_3$ by the rule $(\zeta(z))_{\nu(i,j)} = (\gamma(z_i))_j$ for $z \in \mathcal{E}_c$ and $i,j \in \omega$. It is clear from the definition and the fact that the norm and product topology coincide on the compactum C' that $\zeta \colon Z_c \to Z_3$ is a closed imbedding. Note that $\|\zeta(z)\| = \|z\|$ for each $z \in \mathcal{E}_c$ and hence ζ is also a closed imbedding with respect to the norm topologies (recall that the norm topology is generated by the product topology together with the norm function). We now define an $H \colon \mathcal{E}_c \to \mathcal{H}_V(M)$ by

(10.21) $$H_z(x) = \begin{cases} G^0_{\zeta(z)}(x), & \text{if } x \in V_0; \\ G^k_{\gamma(z_{k-1})}(x), & \text{if } x \in V_k \text{ for some } k \in \mathbb{N}; \\ x, & \text{if } x \in M \setminus V. \end{cases}$$

for $z \in \mathcal{E}_c$. Since the V_k's form a null sequence it is clear that every H_z is a homeomorphism of M and that H_z depends continuously on $z \in \mathcal{E}_c$. Let $\Pi \colon \mathcal{H}_V(M) \to \mathcal{H}_{V_0}(M)$ be the continuous map that is defined by $\Pi(h) = (h{\restriction}V_0) \cup \mathrm{id}_{M \setminus V_0}$. Since ζ and G^0 are closed imbeddings and $\Pi \circ H = G^0 \circ \zeta$ we have by Lemma 10.3 that $H \colon \mathcal{E}_c \to \mathcal{H}_O(M)$ is also a closed imbedding.

Let $k \in \omega$ and let D_k be a countable dense subset of V_k with $D_k \cap \mathbb{R}_k = \emptyset$ and $A_k \subset D_k$. Since P is countable we may assume that $G^k_z(D_k) = D_k$ for each $z \in P$. Let $\mathbb{Q}_3$ be the additive group $\{i3^j \colon i,j \in \mathbb{Z}\}$ and note that $C \cap \mathbb{Q}_3 = Q$. Let $\mathbb{Q}_3^k$ be the copy of $\mathbb{Q}_3$ that lies in $\mathbb{R}_k$. With strong local homogeneity of μ^n we may assume that the set D has the properties

(10.22) $$\begin{aligned} D \cap V_0 &= D_0, \\ D \cap V_k &= D_k \cup \mathbb{Q}_3^k \quad \text{for } k \in \mathbb{N}. \end{aligned}$$

We verify that

(10.23) $$\mathcal{E} = \{z \in \mathcal{E}_c \colon H_z(D) = D\}$$

and hence that $H{\restriction}\mathcal{E}$ is a closed imbedding of $\mathcal{E}$ into $\mathcal{H}_O(M,D)$. If $H_z \in \mathcal{H}(M,D)$ and $k \in \mathbb{N}$, then $H_z(0_k) = \|\gamma(z_{k-1})\| = z_{k-1} \in \mathbb{Q}_3$. Since $z \in \mathcal{E}_c$ we also have $z_{k-1} \in C$ and hence $z_{k-1} \in Q$. Thus $z \in \mathcal{E}$. Consider now a $z \in \mathcal{E}$. If $x \in V_k \setminus \mathbb{R}_k$ for some $k \in \omega$, then by property (c) there is a $z' \in P$ such that $H_z(x) = G^k_{z'}(x)$. Since $G^k_{z'}(D_k) = D_k$ we have that $x \in D_k = D \cap V_k \setminus \mathbb{R}_k$ if and only if $H_z(x) \in D_k$.

Note that $H_z(\mathbb{R}_0) = \mathbb{R}_0$ and that this set is disjoint from D. Consider finally the case $x \in \mathbb{R}_k$ for $k \in \mathbb{N}$. Then $z_{k-1} \in Q \subset \mathbb{Q}_3$ and $H_z(x) = G^k_{\gamma(z_{k-1})}(x) = x + \|\gamma(z_{k-1})\| = x + z_{k-1}$ which is in $\mathbb{Q}_3$ if and only if $x \in \mathbb{Q}_3$.

Consider now the topology $\mathfrak{T}$ that $\mathcal{H}_O(M,D)$ inherits from $D'^{D'}$. Let $\mathfrak{T}'$ be the topology that $\mathcal{H}(M)$ inherits from the product space $M^{D'}$ and note that $\mathfrak{T}'$ restricts to $\mathfrak{T}$ on $\mathcal{H}_O(M,D)$. We first verify that $H\colon Z_c \to (\mathcal{H}(M), \mathfrak{T}')$ is continuous. Let $x \in D'$. If $x \notin V$ or if $x \in V_k$ for some $k \in \mathbb{N}$, then $H_z(x)$ depends on at most a single coordinate of z so continuity with respect to the product topology is obvious. Let $x \in V_0$ and thus $x \in D_0 \subset V_0 \setminus \mathbb{R}_0$. Then by property (c), $G^0_{z'}(x)$ depends on only finitely many coordinates of $z' \in Z_3$ and hence $H_z(x) = G^0_{\zeta(z)}(x)$ depends also on only finitely many coordinates of $z \in Z_c$. This shows that H is continuous with respect to the product topologies. With property (d) we find that $\beta_0 \circ H = \beta_0 \circ G^0 \circ \zeta$ is a closed imbedding of Z_c into $\beta_0(\mathcal{H}(M))$. Since $A_0 \subset D'$ we have that $\beta_0\colon (\mathcal{H}(M), \mathfrak{T}') \to M^{A_0}$ is continuous. Thus with Lemma 10.3 we may conclude that $H\colon Z_c \to (\mathcal{H}(M), \mathfrak{T}')$ is a closed imbedding. Since $Z = H^{-1}(\mathcal{H}_O(M,D))$ we also have that $H{\restriction}Z$ is a closed imbedding of Z in $(\mathcal{H}_O(M,D), \mathfrak{T})$.

Consider the point $0_1 \in \mathbb{Q}_3^1$. For every $a \in D$ we define $Y_a = \{h \in \mathcal{H}_O(M,D): h(0_1) = a\}$ and we note that every Y_a is closed with respect to $\mathfrak{T}$ and that $\bigcup_{a \in D} Y_a = \mathcal{H}_O(M,D)$. Let for $i \in \mathbb{N}$, $z^i = (3^{-i}, 0, 0, \dots) \in \mathcal{E}$ and let $h \in Y_a$. Note that $\lim_{i \to \infty} h \circ H_0^{-1} \circ H_{z^i} = h$ in $\mathcal{H}_O(M,D)$ but $h \circ H_0^{-1} \circ H_{z^i} \notin Y_a$ because $h(H_0^{-1}(H_{z^i}(0_1))) = h((3^{-i})_1) \neq h(0_1) = a$. Thus Y_a is nowhere dense in $\mathcal{H}_O(M,D)$ and condition (5') of Definition 8.7 is satisfied.

Let $\pm\Omega = \pm\infty_0 \in \hat{\mathbb{R}}_0$. By the same argument as we used in the proof of Theorem 10.2 we have for every unbounded $A \subset \mathcal{E}$ that

(10.24) $$\operatorname{diam}_{\hat{\rho}}\{H_z : z \in A\} \geq \rho(-\Omega, \Omega).$$

Finally, we consider the natural stratification of $\mathcal{E}$ that satisfies Definition 8.3. Let $T = Q^{<\omega}$ and for each $s = q_1 \dots q_k \in T$ we put

(10.25) $$\mathcal{E}_s = \{z \in \mathcal{E} : z_{i-1} = q_i \text{ for } 1 \leq i \leq k\}.$$

The remainder of the proof is analogous to latter part of the proof of Theorem 10.2. □

REMARK 10.5. The 'zero-dimensional Menger space' is of course the Cantor set $\mathfrak{C}$ and we showed in [21] that if D is a countable dense subset of $\mathfrak{C}$, then $\mathcal{H}(\mathfrak{C}, D)$ is homeomorphic to $\mathbb{Q}^\omega$.

Let $n \in \mathbb{N}$. A nowhere dense compact subset X of the $(n+1)$-sphere S^{n+1} is called an *n-dimensional Sierpiński carpet* if the collection of components $\{U_i : i \in \mathbb{N}\}$ of $S^{n+1} \setminus X$ forms a null sequence such that the closures of the U_i's are a pairwise disjoint collection and every $S^{n+1} \setminus U_i$ is an $(n+1)$-cell. According to Whyburn [45] (for $n = 1$) and Cannon [13] (for $n \geq 2$) this space is topologically unique if $n \neq 3$ and we will denote this space by M_n^{n+1}.

It is shown in Dijkstra [15, Remarks 3 and 4] that there exist imbeddings of complete Erdős space in $\mathcal{H}(M_n^{n+1})$ that are similar to the ones used in the proof of Theorem 10.4 and one can construct for M_n^{n+1} an argument that is analogous to that proof. A difference is that we have to be careful with the selection of the countable dense set D. This is because M_n^{n+1} is not homogeneous. The result is the following theorem, the proof of which will appear in Dijkstra and Visser [24].

THEOREM 10.6. *Let $n \in \mathbb{N}\setminus\{3\}$, let $\{U_i : i \in \mathbb{N}\}$ be the collection of components of $S^{n+1}\setminus M_n^{n+1}$, and let D be a countable dense subset of M_n^{n+1}. If O is a nonempty open subset of M_n^{n+1} such that either $D \cap \partial U_i = \emptyset$ for every i with $\partial U_i \subset O$ or $D \cap \partial U_i$ is dense in ∂U_i for every i with $\partial U_i \subset O$, then $\mathcal{H}(M_n^{n+1}, D)$ is homeomorphic to Erdős space.*

REMARK 10.7. If X is locally compact but not compact, then we gave $\mathcal{H}(X)$ the topology that is inherited from $\mathcal{H}(\alpha X)$ because the compact-open topology on $\mathcal{H}(X)$ is in general not a group topology. We will now verify that the theorems in this section remain valid if we equip $\mathcal{H}(X)$ with the compact-open topology. Let us denote this space by $\mathcal{H}(X)_{\mathrm{co}}$. Note that multiplication is a continuous operation on $\mathcal{H}(X)_{\mathrm{co}}$, see Arens [6], and that in the proofs in this section we did not use the continuity of the inverse.

Let us first consider Theorem 10.1. This theorem remains valid if M is locally compact and we replace $\mathcal{H}_O(M, D_1)$ by $\mathcal{H}_O(M, D_1)_{\mathrm{co}}$ (seen as a subspace of $\mathcal{H}(M)_{\mathrm{co}}$). In the first part of the proof where $\mathcal{T}$ is shown to be a witness topology it is enough to add the condition that K is compact when defining the subbasis $\mathcal{S}$. To adapt the proof that $\mathcal{T}$ is $F_{\sigma\delta}$ it suffices to let $\{(A_i, B_i) : i \in \mathbb{N}\}$ be a collection of pairs of open subsets of M with disjoint closures such that $M \setminus B_i$ is compact for every $i \in \mathbb{N}$ and for every pair (A, B) of disjoint closed subsets of M such that A is compact there exists an $i \in \mathbb{N}$ with $A \subset A_i$ and $B \subset B_i$.

We now discuss the adaptation of the proofs of Theorems 10.2 and 10.4. We concentrate on the $\mathbb{R}^n$ case – the cases for $\mathfrak{Q}$ and μ^n are completely analogous. We define the open set

(10.26) $$Y = \bigcup \{U : U \text{ is a locally connected open subset of } M\}.$$

Since Y is locally compact and locally connected we have according to Arens [6] that $\mathcal{H}(Y)_{\mathrm{co}} = \mathcal{H}(Y)$. Let ρ be a compatible metric on αY and define the pseudo-metric $\hat\rho$ on $\mathcal{H}(M)_{\mathrm{co}}$ by

(10.27) $$\hat\rho(f, g) = \sup_{x \in Y} \rho(f(x), g(x)),$$

where we noted that $\mathcal{H}(M) = \mathcal{H}(M, Y)$. Since $h \mapsto h\restriction Y$ defines a clearly continuous map from $\mathcal{H}(M)_{\mathrm{co}}$ to $\mathcal{H}(Y)_{\mathrm{co}}$ we have that $\hat\rho$ generates a topology on $\mathcal{H}(M)$ that is coarser than the compact-open topology. We may assume that the compactum $\hat{\mathbb{R}} \times J^{n-1}$ is contained in Y. Consequently, $\hat\rho$ is a compatible metric on $\mathcal{H}_W(M)_{\mathrm{co}}$, where $W = \mathbb{R} \times (-1, 1)^{n-1}$. As a consequence we have that in order to establish the properties of $H : \mathcal{E} \to \mathcal{H}_W(M)_{\mathrm{co}}$ we may use $\hat\rho$ as a metric on the codomain. Note that formula (10.12) remains valid. In verifying that the system $(E_s)_{s \in T}$ satisfies Definition 8.7 we note that $\hat\rho$ is right invariant which means that conditions (2') and (4') still follow from (10.12) by precisely the same argument.

Bibliography

[1] M. Abry and J. J. Dijkstra, *On topological Kadec norms*, Math. Ann. **332** (2005), 759–765. MR2179775 (2006h:54007)
[2] M. Abry and J. J. Dijkstra, *Universal spaces for almost n-dimensionality*, Proc. Amer. Math. Soc. **135** (2007), 725–733. MR2302584 (2008a:54042)
[3] M. Abry, J. J. Dijkstra, and J. van Mill, *Sums of almost zero-dimensional spaces*, Topology Proc. **29** (2005), 1–12. MR2182913 (2006g:54040)
[4] M. Abry, J. J. Dijkstra, and J. van Mill, *On one-point connectifications*, Topology Appl. **154** (2007), 725–733. MR2280916 (2008c:54033)
[5] R. D. Anderson, *Spaces of homeomorphisms of finite graphs*, unpublished manuscript.
[6] R. Arens, *Topologies for homeomorphism groups*, Amer. J. Math. **68** (1946), 593–610. MR0019916 (8:479i)
[7] R. Bennett, *Countable dense homogeneous spaces*, Fund. Math. **74** (1972), 189–194. MR0301711 (46:866)
[8] C. Bessaga and A. Pełczyński, *Selected Topics in Infinite-Dimensional Topology*, Mathematical Monographs, Vol. 58, PWN—Polish Scientific Publishers, Warsaw, 1975. MR0478168 (57:17657)
[9] M. Bestvina, *Characterizing k-dimensional universal Menger compacta*, Mem. Amer. Math. Soc. **71** (1988), no. 380. MR920964 (89g:54083)
[10] L. E. J. Brouwer, *On the structure of perfect sets of points*, Proc. Akad. Amsterdam **12** (1910), 785–794.
[11] L. E. J. Brouwer, *Some remarks on the coherence type η*, Proc. Akad. Amsterdam **15** (1913), 1256–1263.
[12] W. D. Bula and L. G. Oversteegen, *A characterization of smooth Cantor bouquets*, Proc. Amer. Math. Soc. **108** (1990), 529–534. MR991691 (90d:54066)
[13] J. W. Cannon, *A positional characterization of the $(n-1)$-dimensional Sierpiński curve in $S^n (n \neq 4)$*, Fund. Math. **79** (1973), 107–112. MR0319203 (47:7748)
[14] W. J. Charatonik, *The Lelek fan is unique*, Houston J. Math. **15** (1989), 27–34. MR1002079 (90f:54050)
[15] J. J. Dijkstra, *On homeomorphism groups of Menger continua*, Trans. Amer. Math. Soc. **357** (2005), 2665-2679. MR2139522 (2006c:57031)
[16] J. J. Dijkstra, *A criterion for Erdős spaces*, Proc. Edinburgh Math. Soc. **48** (2005), 595–601. MR2171187 (2006g:54041)
[17] J. J. Dijkstra, *On homeomorphism groups and the compact-open topology*, Amer. Math. Monthly **112** (2005), 910–912. MR2186833 (2006h:54033)
[18] J. J. Dijkstra, *A homogeneous space that is one-dimensional but not cohesive*, Houston J. Math. **32** (2006), 1093–1099. MR2268471 (2008i:54031)
[19] J. J. Dijkstra, *Characterizing stable complete Erdős space*, Israel J. Math, in press.
[20] J. J. Dijkstra and J. van Mill, *Homeomorphism groups of manifolds and Erdős space*, Electron. Res. Announc. Amer. Math. Soc. **10** (2004), 29–38. MR2048429 (2005c:57025)
[21] J. J. Dijkstra and J. van Mill, *On the group of homeomorphisms of the real line that map the pseudoboundary onto itself*, Canad. J. Math. **58** (2006), 529–547. MR2223455 (2007b:57064)
[22] J. J. Dijkstra and J. van Mill, *Characterizing complete Erdős space*, Canad J. Math. **61** (2009), 124–140. MR2488452
[23] J. J. Dijkstra, J. van Mill, and J. Steprāns, *Complete Erdős space is unstable*, Math. Proc. Cambridge Philos. Soc. **137** (2004), 465-473. MR2092071 (2005g:54076)
[24] J. J. Dijkstra and D. Visser, *Homeomorphism groups of Sierpiński carpets and Erdős space*, Fund. Math. **207** (2010), 1–19.

[25] T. Dobrowolski, J. Grabowski, and K. Kawamura, *Topological type of weakly closed subgroups in Banach spaces*, Studia Math. **118** (1996), 49–62. MR1373624 (97d:46013)

[26] T. Dobrowolski and H. Toruńczyk, *Separable complete ANRs admitting a group structure are Hilbert manifolds*, Topology Appl. **12** (1981), 229–235. MR623731 (83a:58007)

[27] R. Engelking, *Dimension Theory*, North-Holland, Amsterdam, 1978. MR0482697 (58:2753b)

[28] A. J. M. van Engelen, *Homogeneous Zero-Dimensional Absolute Borel Sets*, CWI Tract, Vol. 27, Centre for Mathematics and Computer Science, Amsterdam, 1986. MR851765 (87j:54058)

[29] P. Erdős, *The dimension of the rational points in Hilbert space*, Ann. of Math. (2) **41** (1940), 734–736. MR0003191 (2:178a)

[30] S. Ferry, *The homeomorphism group of a compact Hilbert cube manifold is an* ANR, Ann. of Math. (2) **106** (1977), 101–119. MR0461536 (57:1521)

[31] K. Kawamura, L. G. Oversteegen, and E. D. Tymchatyn, *On homogeneous totally disconnected 1-dimensional spaces*, Fund. Math. **150** (1996), 97–112. MR1391294 (97d:54060)

[32] J. E. Keesling, *Using flows to construct Hilbert space factors of function spaces*, Trans. Amer. Math. Soc. **161** (1971), 1–24. MR0283751 (44:981)

[33] A. Lelek, *On plane dendroids and their end points in the classical sense*, Fund. Math. **49** (1960), 301–319. MR0133806 (24:A3631)

[34] M. Levin and R. Pol, *A metric condition which implies dimension ≤ 1*, Proc. Amer. Math. Soc. **125** (1997), 269–273. MR1389528 (97e:54033)

[35] R. Luke and W. K. Mason, *The space of homeomorphisms of a compact two-manifold is an Absolute Neighborhood Retract*, Trans. Amer. Math. Soc. **164** (1972), 275–285. MR0301693 (46:849)

[36] J. C. Mayer and L. G. Oversteegen. *A topological characterization of $\mathbb{R}$-trees*, Trans. Amer. Math. Soc. **320** (1990), 395–415. MR961626 (90k:54031)

[37] J. van Mill, *The Infinite-Dimensional Topology of Function Spaces*, North-Holland Publishing Co., Amsterdam, 2001. MR1851014 (2002h:57031)

[38] L. G. Oversteegen and E. D. Tymchatyn, *On the dimension of certain totally disconnected spaces*, Proc. Amer. Math. Soc. **122** (1994), 885–891. MR1273515 (95b:54040)

[39] R. Pol, *There is no universal totally disconnected space*, Fund. Math. **70** (1973), 265–267. MR0322778 (48:1139)

[40] W. Sierpiński, *Sur une définition topologique des ensembles $F_{\sigma\delta}$*, Fund. Math. **6** (1924), 24–29.

[41] J. R. Steel, *Analytic sets and Borel isomorphisms*, Fund. Math. **108** (1980), 83–88. MR594307 (82b:03091)

[42] H. Toruńczyk, *Homeomorphism groups of compact Hilbert cube manifolds which are manifolds*, Bull. Acad. Polon. Sci. Sér. Sci. Math. Astronom. Phys. **25** (1977), 401–408. MR0515792 (58:24302)

[43] H. Toruńczyk, *On CE-images of the Hilbert cube and characterizations of Q–manifolds*, Fund. Math. **106** (1980), 31–40. MR585543 (83g:57006)

[44] H. Toruńczyk, *Characterizing Hilbert space topology*, Fund. Math. **111** (1981), 247–262. MR611763 (82i:57016)

[45] G. T. Whyburn, *Topological characterization of the Sierpiński curve*, Fund. Math. **45** (1958), 320–324. MR0099638 (20:6077)

Editorial Information

To be published in the *Memoirs*, a paper must be correct, new, nontrivial, and significant. Further, it must be well written and of interest to a substantial number of mathematicians. Piecemeal results, such as an inconclusive step toward an unproved major theorem or a minor variation on a known result, are in general not acceptable for publication.

Papers appearing in *Memoirs* are generally at least 80 and not more than 200 published pages in length. Papers less than 80 or more than 200 published pages require the approval of the Managing Editor of the Transactions/Memoirs Editorial Board. Published pages are the same size as those generated in the style files provided for $\mathcal{AMS}$-LaTeX or $\mathcal{AMS}$-TeX.

Information on the backlog for this journal can be found on the AMS website starting from http://www.ams.org/memo.

A Consent to Publish and Copyright Agreement is required before a paper will be published in the *Memoirs*. After a paper is accepted for publication, the Providence office will send a Consent to Publish and Copyright Agreement to all authors of the paper. By submitting a paper to the *Memoirs*, authors certify that the results have not been submitted to nor are they under consideration for publication by another journal, conference proceedings, or similar publication.

Information for Authors

Memoirs is an author-prepared publication. Once formatted for print and on-line publication, articles will be published as is with the addition of AMS-prepared frontmatter and backmatter. Articles are not copyedited; however, confirmation copy will be sent to the authors.

Initial submission. The AMS uses Centralized Manuscript Processing for initial submissions. Authors should submit a PDF file using the Initial Manuscript Submission form found at www.ams.org/submission/memo, or send one copy of the manuscript to the following address: Centralized Manuscript Processing, MEMOIRS OF THE AMS, 201 Charles Street, Providence, RI 02904-2294 USA. If a paper copy is being forwarded to the AMS, indicate that it is for *Memoirs* and include the name of the corresponding author, contact information such as email address or mailing address, and the name of an appropriate Editor to review the paper (see the list of Editors below).

The paper must contain a *descriptive title* and an *abstract* that summarizes the article in language suitable for workers in the general field (algebra, analysis, etc.). The *descriptive title* should be short, but informative; useless or vague phrases such as "some remarks about" or "concerning" should be avoided. The *abstract* should be at least one complete sentence, and at most 300 words. Included with the footnotes to the paper should be the 2010 *Mathematics Subject Classification* representing the primary and secondary subjects of the article. The classifications are accessible from www.ams.org/msc/. The Mathematics Subject Classification footnote may be followed by a list of *key words and phrases* describing the subject matter of the article and taken from it. Journal abbreviations used in bibliographies are listed in the latest *Mathematical Reviews* annual index. The series abbreviations are also accessible from www.ams.org/msnhtml/serials.pdf. To help in preparing and verifying references, the AMS offers MR Lookup, a Reference Tool for Linking, at www.ams.org/mrlookup/.

Electronically prepared manuscripts. The AMS encourages electronically prepared manuscripts, with a strong preference for $\mathcal{AMS}$-LaTeX. To this end, the Society has prepared $\mathcal{AMS}$-LaTeX author packages for each AMS publication. Author packages include instructions for preparing electronic manuscripts, samples, and a style file that generates the particular design specifications of that publication series. Though $\mathcal{AMS}$-LaTeX is the highly preferred format of TeX, author packages are also available in $\mathcal{AMS}$-TeX.

Authors may retrieve an author package for *Memoirs of the AMS* from www.ams.org/journals/memo/memoauthorpac.html or via FTP to ftp.ams.org (login as anonymous, enter your complete email address as password, and type cd pub/author-info). The

AMS Author Handbook and the *Instruction Manual* are available in PDF format from the author package link. The author package can also be obtained free of charge by sending email to `tech-support@ams.org` or from the Publication Division, American Mathematical Society, 201 Charles St., Providence, RI 02904-2294, USA. When requesting an author package, please specify $\mathcal{AMS}$-LaTeX or $\mathcal{AMS}$-TeX and the publication in which your paper will appear. Please be sure to include your complete mailing address.

After acceptance. The source files for the final version of the electronic manuscript should be sent to the Providence office immediately after the paper has been accepted for publication. The author should also submit a PDF of the final version of the paper to the editor, who will forward a copy to the Providence office.

Accepted electronically prepared files can be submitted via the web at `www.ams.org/submit-book-journal/`, sent via FTP, or sent on CD to the Electronic Prepress Department, American Mathematical Society, 201 Charles Street, Providence, RI 02904-2294 USA. TeX source files and graphic files can be transferred over the Internet by FTP to the Internet node `ftp.ams.org` (130.44.1.100). When sending a manuscript electronically via CD, please be sure to include a message indicating that the paper is for the *Memoirs*.

Electronic graphics. Comprehensive instructions on preparing graphics are available at `www.ams.org/authors/journals.html`. A few of the major requirements are given here.

Submit files for graphics as EPS (Encapsulated PostScript) files. This includes graphics originated via a graphics application as well as scanned photographs or other computer-generated images. If this is not possible, TIFF files are acceptable as long as they can be opened in Adobe Photoshop or Illustrator.

Authors using graphics packages for the creation of electronic art should also avoid the use of any lines thinner than 0.5 points in width. Many graphics packages allow the user to specify a "hairline" for a very thin line. Hairlines often look acceptable when proofed on a typical laser printer. However, when produced on a high-resolution laser imagesetter, hairlines become nearly invisible and will be lost entirely in the final printing process.

Screens should be set to values between 15% and 85%. Screens which fall outside of this range are too light or too dark to print correctly. Variations of screens within a graphic should be no less than 10%.

Inquiries. Any inquiries concerning a paper that has been accepted for publication should be sent to `memo-query@ams.org` or directly to the Electronic Prepress Department, American Mathematical Society, 201 Charles St., Providence, RI 02904-2294 USA.

Editors

This journal is designed particularly for long research papers, normally at least 80 pages in length, and groups of cognate papers in pure and applied mathematics. Papers intended for publication in the *Memoirs* should be addressed to one of the following editors. The AMS uses Centralized Manuscript Processing for initial submissions to AMS journals. Authors should follow instructions listed on the Initial Submission page found at www.ams.org/memo/memosubmit.html.

Algebra, to ALEXANDER KLESHCHEV, Department of Mathematics, University of Oregon, Eugene, OR 97403-1222; e-mail: ams@noether.uoregon.edu

Algebraic geometry, to DAN ABRAMOVICH, Department of Mathematics, Brown University, Box 1917, Providence, RI 02912; e-mail: amsedit@math.brown.edu

Algebraic geometry and its applications, to MINA TEICHER, Emmy Noether Research Institute for Mathematics, Bar-Ilan University, Ramat-Gan 52900, Israel; e-mail: teicher@macs.biu.ac.il

Algebraic topology, to ALEJANDRO ADEM, Department of Mathematics, University of British Columbia, Room 121, 1984 Mathematics Road, Vancouver, British Columbia, Canada V6T 1Z2; e-mail: adem@math.ubc.ca

Combinatorics, to JOHN R. STEMBRIDGE, Department of Mathematics, University of Michigan, Ann Arbor, Michigan 48109-1109; e-mail: JRS@umich.edu

Commutative and homological algebra, to LUCHEZAR L. AVRAMOV, Department of Mathematics, University of Nebraska, Lincoln, NE 68588-0130; e-mail: avramov@math.unl.edu

Complex analysis and harmonic analysis, to ALEXANDER NAGEL, Department of Mathematics, University of Wisconsin, 480 Lincoln Drive, Madison, WI 53706-1313; e-mail: nagel@math.wisc.edu

Differential geometry and global analysis, to CHRIS WOODWARD, Department of Mathematics, Rutgers University, 110 Frelinghuysen Road, Piscataway, NJ 08854; e-mail: ctw@math.rutgers.edu

Dynamical systems and ergodic theory and complex analysis, to YUNPING JIANG, Department of Mathematics, CUNY Queens College and Graduate Center, 65-30 Kissena Blvd., Flushing, NY 11367; e-mail: Yunping.Jiang@qc.cuny.edu

Functional analysis and operator algebras, to DIMITRI SHLYAKHTENKO, Department of Mathematics, University of California, Los Angeles, CA 90095; e-mail: shlyakht@math.ucla.edu

Geometric analysis, to WILLIAM P. MINICOZZI II, Department of Mathematics, Johns Hopkins University, 3400 N. Charles St., Baltimore, MD 21218; e-mail: trans@math.jhu.edu

Geometric topology, to MARK FEIGHN, Math Department, Rutgers University, Newark, NJ 07102; e-mail: feighn@andromeda.rutgers.edu

Harmonic analysis, representation theory, and Lie theory, to E. P. VAN DEN BAN, Department of Mathematics, Utrecht University, P.O. Box 80 010, 3508 TA Utrecht, The Netherlands; e-mail: E.P.vandenBan@uu.nl

Logic, to STEFFEN LEMPP, Department of Mathematics, University of Wisconsin, 480 Lincoln Drive, Madison, Wisconsin 53706-1388; e-mail: lempp@math.wisc.edu

Number theory, to JONATHAN ROGAWSKI, Department of Mathematics, University of California, Los Angeles, CA 90095; e-mail: jonr@math.ucla.edu

Number theory, to SHANKAR SEN, Department of Mathematics, 505 Malott Hall, Cornell University, Ithaca, NY 14853; e-mail: ss70@cornell.edu

Partial differential equations, to GUSTAVO PONCE, Department of Mathematics, South Hall, Room 6607, University of California, Santa Barbara, CA 93106; e-mail: ponce@math.ucsb.edu

Partial differential equations and dynamical systems, to PETER POLACIK, School of Mathematics, University of Minnesota, Minneapolis, MN 55455; e-mail: polacik@math.umn.edu

Probability and statistics, to RICHARD BASS, Department of Mathematics, University of Connecticut, Storrs, CT 06269-3009; e-mail: bass@math.uconn.edu

Real analysis and partial differential equations, to DANIEL TATARU, Department of Mathematics, University of California, Berkeley, Berkeley, CA 94720; e-mail: tataru@math.berkeley.edu

All other communications to the editors, should be addressed to the Managing Editor, ROBERT GURALNICK, Department of Mathematics, University of Southern California, Los Angeles, CA 90089-1113; e-mail: guralnic@math.usc.edu.

Titles in This Series

981 **Dillon Mayhew, Gordon Royle, and Geoff Whittle,** The internally 4-connected binary matroids with no $M(K_{3,3})$-Minor, 2010

980 **Liviu I. Nicolaescu,** Tame flows, 2010

979 **Jan J. Dijkstra and Jan van Mill,** Erdős space and homeomorphism groups of manifolds, 2010

978 **Gilles Pisier,** Complex interpolation between Hilbert, Banach and operator spaces, 2010

977 **Thomas Lam, Luc Lapointe, Jennifer Morse, and Mark Shimozono,** Affine insertion and Pieri rules for the affine Grassmannian, 2010

976 **Alfonso Castro and Víctor Padrón,** Classification of radial solutions arising in the study of thermal structures with thermal equilibrium or no flux at the boundary, 2010

975 **Javier Ribón,** Topological classification of families of diffeomorphisms without small divisors, 2010

974 **Pascal Lefèvre, Daniel Li, Hervé Queffélec, and Luis Rodríguez-Piazza,** Composition operators on Hardy-Orlicz space, 2010

973 **Peter O'Sullivan,** The generalised Jacobson-Morosov theorem, 2010

972 **Patrick Iglesias-Zemmour,** The moment maps in diffeology, 2010

971 **Mark D. Hamilton,** Locally toric manifolds and singular Bohr-Sommerfeld leaves, 2010

970 **Klaus Thomsen,** C^*-algebras of homoclinic and heteroclinic structure in expansive dynamics, 2010

969 **Makoto Sakai,** Small modifications of quadrature domains, 2010

968 **L. Nguyen Van Thé,** Structural Ramsey theory of metric spaces and topological dynamics of isometry groups, 2010

967 **Zeng Lian and Kening Lu,** Lyapunov exponents and invariant manifolds for random dynamical systems in a Banach space, 2010

966 **H. G. Dales, A. T.-M. Lau, and D. Strauss,** Banach algebras on semigroups and on their compactifications, 2010

965 **Michael Lacey and Xiaochun Li,** On a conjecture of E. M. Stein on the Hilbert transform on vector fields, 2010

964 **Gelu Popescu,** Operator theory on noncommutative domains, 2010

963 **Huaxin Lin,** Approximate homotopy of homomorphisms from $C(X)$ into a simple C^*-algebra, 2010

962 **Adam Coffman,** Unfolding CR singularities, 2010

961 **Marco Bramanti, Luca Brandolini, Ermanno Lanconelli, and Francesco Uguzzoni,** Non-divergence equations structured on Hörmander vector fields: Heat kernels and Harnack inequalities, 2010

960 **Olivier Alvarez and Martino Bardi,** Ergodicity, stabilization, and singular perturbations for Bellman-Isaacs equations, 2010

959 **Alvaro Pelayo,** Symplectic actions of 2-tori on 4-manifolds, 2010

958 **Mark Behrens and Tyler Lawson,** Topological automorphic forms, 2010

957 **Ping-Shun Chan,** Invariant representations of GSp(2) under tensor product with a quadratic character, 2010

956 **Richard Montgomery and Michail Zhitomirskii,** Points and curves in the Monster tower, 2010

955 **Martin R. Bridson and Daniel Groves,** The quadratic isoperimetric inequality for mapping tori of free group automorphisms, 2010

954 **Volker Mayer and Mariusz Urbański,** Thermodynamical formalism and multifractal analysis for meromorphic functions of finite order, 2010

For a complete list of titles in this series, visit the AMS Bookstore at **www.ams.org/bookstore/**.

Titles in This Series

909 **Cameron McA. Gordon and Ying-Qing Wu,** Toroidal Dehn fillings on hyperbolic 3-manifolds, 2008

908 **J.-L. Waldspurger,** L'endoscopie tordue n'est pas si tordue, 2008

907 **Yuanhua Wang and Fei Xu,** Spinor genera in characteristic 2, 2008

906 **Raphaël S. Ponge,** Heisenberg calculus and spectral theory of hypoelliptic operators on Heisenberg manifolds, 2008

905 **Dominic Verity,** Complicial sets characterising the simplicial nerves of strict ω-categories, 2008

904 **William M. Goldman and Eugene Z. Xia,** Rank one Higgs bundles and representations of fundamental groups of Riemann surfaces, 2008

903 **Gail Letzter,** Invariant differential operators for quantum symmetric spaces, 2008

902 **Bertrand Toën and Gabriele Vezzosi,** Homotopical algebraic geometry II: Geometric stacks and applications, 2008

901 **Ron Donagi and Tony Pantev (with an appendix by Dmitry Arinkin),** Torus fibrations, gerbes, and duality, 2008

900 **Wolfgang Bertram,** Differential geometry, Lie groups and symmetric spaces over general base fields and rings, 2008

899 **Piotr Hajłasz, Tadeusz Iwaniec, Jan Malý, and Jani Onninen,** Weakly differentiable mappings between manifolds, 2008

898 **John Rognes,** Galois extensions of structured ring spectra/Stably dualizable groups, 2008

897 **Michael I. Ganzburg,** Limit theorems of polynomial approximation with exponential weights, 2008

896 **Michael Kapovich, Bernhard Leeb, and John J. Millson,** The generalized triangle inequalities in symmetric spaces and buildings with applications to algebra, 2008

895 **Steffen Roch,** Finite sections of band-dominated operators, 2008

894 **Martin Dindoš,** Hardy spaces and potential theory on C^1 domains in Riemannian manifolds, 2008

893 **Tadeusz Iwaniec and Gaven Martin,** The Beltrami Equation, 2008

892 **Jim Agler, John Harland, and Benjamin J. Raphael,** Classical function theory, operator dilation theory, and machine computation on multiply-connected domains, 2008

891 **John H. Hubbard and Peter Papadopol,** Newton's method applied to two quadratic equations in $\mathbb{C}^2$ viewed as a global dynamical system, 2008

890 **Steven Dale Cutkosky,** Toroidalization of dominant morphisms of 3-folds, 2007

889 **Michael Sever,** Distribution solutions of nonlinear systems of conservation laws, 2007

888 **Roger Chalkley,** Basic global relative invariants for nonlinear differential equations, 2007

887 **Charlotte Wahl,** Noncommutative Maslov index and eta-forms, 2007

886 **Robert M. Guralnick and John Shareshian,** Symmetric and alternating groups as monodromy groups of Riemann surfaces I: Generic covers and covers with many branch points, 2007

885 **Jae Choon Cha,** The structure of the rational concordance group of knots, 2007

884 **Dan Haran, Moshe Jarden, and Florian Pop,** Projective group structures as absolute Galois structures with block approximation, 2007

883 **Apostolos Beligiannis and Idun Reiten,** Homological and homotopical aspects of torsion theories, 2007

882 **Lars Inge Hedberg and Yuri Netrusov,** An axiomatic approach to function spaces, spec tral synthesis and Luzin approximation, 2007

881 **Tao Mei,** Operator valued Hardy spaces, 2007

TITLES IN THIS SERIES

880 **Bruce C. Berndt, Geumlan Choi, Youn-Seo Choi, Heekyoung Hahn, Boon Pin Yeap, Ae Ja Yee, Hamza Yesilyurt, and Jinhee Yi,** Ramanujan's forty identities for Rogers-Ramanujan functions, 2007

879 **O. García-Prada, P. B. Gothen, and V. Muñoz,** Betti numbers of the moduli space of rank 3 parabolic Higgs bundles, 2007

878 **Alessandra Celletti and Luigi Chierchia,** KAM stability and celestial mechanics, 2007

877 **María J. Carro, José A. Raposo, and Javier Soria,** Recent developments in the theory of Lorentz spaces and weighted inequalities, 2007

876 **Gabriel Debs and Jean Saint Raymond,** Borel liftings of Borel sets: Some decidable and undecidable statements, 2007

875 **C. Krattenthaler and T. Rivoal,** Hypergéométrie et fonction zêta de Riemann, 2007

874 **Sonia Natale,** Semisolvability of semisimple Hopf algebras of low dimension, 2007

873 **A. J. Duncan,** Exponential genus problems in one-relator products of groups, 2007

872 **Anthony V. Geramita, Tadahito Harima, Juan C. Migliore, and Yong Su Shin,** The Hilbert function of a level algebra, 2007

871 **Pascal Auscher,** On necessary and sufficient conditions for L^p-estimates of Riesz transforms associated to elliptic operators on $\mathbb{R}^n$ and related estimates, 2007

870 **Takuro Mochizuki,** Asymptotic behaviour of tame harmonic bundles and an application to pure twistor D-modules, Part 2, 2007

869 **Takuro Mochizuki,** Asymptotic behaviour of tame harmonic bundles and an application to pure twistor D-modules, Part 1, 2007

868 **Gelu Popescu,** Entropy and multivariable interpolation, 2006

867 **Vilmos Totik,** Metric properties of harmonic measures, 2006

866 **William Craig,** Semigroups underlying first-order logic, 2006

865 **Nathanial P. Brown,** Invariant means and finite representation theory of $C*$-algebras, 2006

864 **John M. Lee,** Fredholm operators and Einstein metrics on conformally compact manifolds, 2006

863 **M. Lübke and A. Teleman,** The Universal Kobayashi-Hitchin correspondence on Hermitian manifolds, 2006

862 **Alberto Canonaco,** The Beilinson complex and canonical rings of irregular surfaces, 2006

861 **Leon A. Takhtajan and Lee-Peng Teo,** Weil-Petersson metric on the universal Teichmüller space, 2006

860 **Thomas M. Fiore,** Pseudo limits, biadjoints and pseudo algebras: Categorical foundations of conformal field theory, 2006

859 **N. Arcozzi, R. Rochberg, and E. Sawyer,** Carleson measures and interpolating sequences for Besov spaces on complex balls, 2006

858 **Enrico Valdinoci, Berardino Sciunzi, and Vasile Ovidiu Savin,** Flat level set regularity of p-Laplace phase transitions, 2006

857 **Donatella Danielli, Nocola Garofalo, and Duy-Minh Nhieu,** Non-doubling Ahlfors measures, perimeter measures, and the characterization of the trace spaces of Sobolev functions in Carnot-Carathéodory spaces, 2006

856 **Vladimir Bolotnikov and Harry Dym,** On boundary interpolation for matrix valued Schur functions, 2006

855 **Yevgenia Kashina, Yorck Sommerhäuser, and Yongchang Zhu,** On higher Frobenius-Schur indicators, 2006

For a complete list of titles in this series, visit the
AMS Bookstore at **www.ams.org/bookstore/**.